AF595042

Research on the Regulatory Mechanism of Algae Reproduction under Abiotic Stress Conditions

Research on the Regulatory Mechanism of Algae Reproduction under Abiotic Stress Conditions

Editor

Koji Mikami

MDPI • Basel • Beijing • Wuhan • Barcelona • Belgrade • Manchester • Tokyo • Cluj • Tianjin

Editor
Koji Mikami
Department of Integrative
Studies of Plant and Animal
Production, School of Food
Industrial Sciences
Miyagi University
Sendai
Japan

Editorial Office
MDPI
St. Alban-Anlage 66
4052 Basel, Switzerland

This is a reprint of articles from the Special Issue published online in the open access journal *Plants* (ISSN 2223-7747) (available at: www.mdpi.com/journal/plants/special_issues/algae_reproduction).

For citation purposes, cite each article independently as indicated on the article page online and as indicated below:

LastName, A.A.; LastName, B.B.; LastName, C.C. Article Title. *Journal Name* **Year**, *Volume Number*, Page Range.

ISBN 978-3-0365-3392-6 (Hbk)
ISBN 978-3-0365-3391-9 (PDF)

Cover image courtesy of Koji Mikami

© 2022 by the authors. Articles in this book are Open Access and distributed under the Creative Commons Attribution (CC BY) license, which allows users to download, copy and build upon published articles, as long as the author and publisher are properly credited, which ensures maximum dissemination and a wider impact of our publications.
The book as a whole is distributed by MDPI under the terms and conditions of the Creative Commons license CC BY-NC-ND.

Contents

About the Editor

Koji Mikami

Dr. Koji Mikami is a professor at the Department of Integrative Studies of Plant and Animal Production in School of Food Industrial Sciences of the Miyagi University, Sendai, Japan from October 1, 2020. Before 2006, his studied development and environmental stress response in terrestrial plants, like angiosperm *Arabidopsis thaliana* and moss *Physcomitrium patens*, most of which were done at the National Institute for Basic Biology, Okazaki, Japan, University of California, Berkeley, USA, and Free University of Berlin, Germany. After joining to the Faculty of Fisheries Sciences, Hokkaido University, Hakodate, Japan from 2006, his main interests changed to physiological and molecular biological researches on marine plants, especially red seaweeds. His efforts have paid for elucidation of regulatory mechanisms of diploid-haploid life cycle and also for development of reverse-genetic experimental methods. Additional fields of interest are the mechanisms regulating response to environmental stress and acquisition of stress tolerance in red seaweeds. His research activity is summarized in, for instance, more than 80 papers in peer-reviewed international journals and more than 10 chapters in books. He has published two books as an editor. He is editorial board member of several international journals and has edited several Special Issues.

Preface to "Research on the Regulatory Mechanism of Algae Reproduction under Abiotic Stress Conditions"

Algae living in the hydrosphere are usually exposed to changes in environmental conditions such as abiotic stresses. Thus, acclimation and acquisition of tolerance to these stresses are indispensable for their sustainable survival. Recent omics analyses represent the primary importance of stress-inducible and repressive expression of genes involved in the biosynthesis and metabolism of intracellular components in adaptive response to abiotic stresses in algae. In addition, physiological research has indicated that transition from growth to reproductive phases is sometimes promoted by abiotic stresses, which promote both sexual and asexual reproductive processes. These are usually expressed as changes in life cycle generation, also known as life cycle trade-off. This Special Issue highlights novel findings that significantly contribute to the development of our understanding of how abiotic stress-inducible reproduction is regulated by physiological responses including the life cycle trade-off. The editor gives thanks to all of the authors of manuscripts and staff of the MDPI management team for their contribution and assistance, respectively, in the preparation of this book.

Koji Mikami
Editor

Editorial

Research on the Regulatory Mechanism of Algae Reproduction under Abiotic Stress Conditions

Koji Mikami 

Department of Integrative Studies of Plant and Animal Production, School of Food Industrial Sciences, Miyagi University, 2-2-1 Hatatate, Taihaku-ku, Sendai 982-0215, Japan; mikamik@myu.ac.jp

The intertidal and subtidal zones are characterized by daily and seasonal fluctuations in environmental conditions. Seaweeds that inhabit these environments face wide-ranging temperatures, nutrient deficiency, changes in salinity, and long periods of desiccation [1–7]. Accordingly, like terrestrial plants, these seaweeds possess the innate ability to acclimate to environmental stresses [8–12].

The 'life-cycle trade-off' is a well-described phenomenon of both algae and terrestrial plants that controls the timing of growth and reproduction in response to environmental stresses; this trade-off can optimize survival by selecting sexual or asexual propagation to promote adaptation to changes in environmental conditions [13–16]. Despite these similarities, the effects of environmental stresses differ between algae and terrestrial plants. For instance, although heat stress negatively affects reproduction in terrestrial plants [17–19], positive effects of heat stress on the sexual life-cycle progression have been observed in sessile red algae of the order Bangiales [13,20] and in the green alga *Volvox carteri* [21]. Thus, elucidation of the regulatory mechanisms of the life-cycle trade-off in seaweeds could provide insights valuable not only for enhancing production during mariculture farming of economically important marine resources but also for sustaining the sea environment via maintenance of seaweed forests.

In *'Bangia'* sp. ESS1 (Bangiales), the asexual life-cycle—which involves the production of asexual spores from thalli—is promoted by heat stress [22]. Moreover, non-lethal temperature stress promotes heat stress tolerance in *'Bangia'* sp. ESS1, which enables survival under otherwise lethal heat stress conditions [23]. Since spore release was observed to coincide with acquisition of tolerance in *'Bangia'* sp. ESS1 [22,23], the promotion of the asexual life-cycle by heat stress is proposed to be triggered by establishment of heat stress tolerance. No spore release was observed in *Neopyropia yezoensis*, a major cultivar of nori in Japan, under heat stress conditions [24]; therefore, it is possible that such promotion of the asexual life-cycle by heat stress and its relationship to the acquisition of heat stress tolerance is genus or species specific in Bangiales and other algae.

This Special Issue on "Research on the Regulatory Mechanism of Algae Reproduction under Abiotic Stress Conditions" comprises five studies covering aspects of the life-cycle trade-off. They address the effects of the loss of water current on stimulation of asexual life-cycle progression [25], the effects of combined heat and nutritional depletion stresses on promotion of the asexual life-cycle [26], the relationship between heat stress tolerance and loss of life-cycle trade-off ability [27], and the role of a heat stress-insensitive asexual life-cycle trade-off in the maintenance of vegetative growth [28]. In addition, Khoa et al. [29] focus on the intrinsic ability to acquire tolerance to lethal heat stress in different *Bangia* species based on memory of non-lethal heat stress in relation to asexual spore release. Thus, the studies in this Special Issue cover a broad range of recent findings on environmental stress-dependent life-cycle trade-offs in seaweeds. In this Editorial, I summarize the highlights of each study and focus on the promotion of the asexual life-cycle under combined environmental stress conditions.

Citation: Mikami, K. Research on the Regulatory Mechanism of Algae Reproduction under Abiotic Stress Conditions. *Plants* **2022**, *11*, 525. https://doi.org/10.3390/plants11040525

Received: 27 January 2022
Accepted: 12 February 2022
Published: 15 February 2022

Publisher's Note: MDPI stays neutral with regard to jurisdictional claims in published maps and institutional affiliations.

Copyright: © 2022 by the author. Licensee MDPI, Basel, Switzerland. This article is an open access article distributed under the terms and conditions of the Creative Commons Attribution (CC BY) license (https://creativecommons.org/licenses/by/4.0/).

Omuro et al. [25] explore the acquisition of freezing tolerance and the promotion of the freezing-dependent asexual life-cycle by loss of hydrodynamic stress in *'Bangia'* sp. ESS1. Since Bangiales inhabit the intertidal zone with dynamic water currents, it is likely that hydrodynamic stress is required for their growth and survival. Membrane fatty acids of *'Bangia'* sp. ESS1 were unsaturated during static culture (lacking water current), which resulted in acquisition of freezing tolerance and asexual spore release after thawing. Though the relationship between asexual spore release and unsaturation of membrane fatty acids needs to be elucidated, there is clearly a tight relationship between freezing tolerance acquisition upon loss of hydrodynamic stress and the promotion of the life-cycle trade-off in *'Bangia'* sp. ESS1.

According to a recent revision of the Bangiales phylogeny, the genus *Bangia sensu lato* was separated into four genera, *Bangia*, *'Bangia'* 1, *'Bangia'* 2, and *'Bangia'* 3 [30]. *'Bangia'* sp. ESS1 belongs to *'Bangia'* 2 [31] and has an intrinsic ability for heat stress memory to acquire heat stress tolerance with the release of asexual spores [22,23]. However, little is known about whether other *Bangia* species also remember and adapt to heat stress. Khoa et al. [29] classified *'Bangia'* sp. ESS2 as *'Bangia'* 3 and compared its heat stress response with those in *'Bangia'* sp. ESS1 and *Bangia atropurpurea* [32]. *'Bangia'* sp. ESS2 was not able to acquire heat stress tolerance and remember previous heat stress, whereas the acquisition of heat stress tolerance but not heat stress memory was observed in *B. atropurpurea*. In addition, the asexual life-cycle was repressed by heat stress in *'Bangia'* sp. ESS2, and *B. atropurpurea* did not release asexual spores under heat stress conditions. Thus, intrinsic heat stress responses, including the life-cycle trade-off, appear to be species-specific. Overall, these findings underscore that there is a relationship between heat stress memory and heat stress-dependent promotion of the asexual life-cycle.

Endo et al. [26] demonstrate a high tolerance of holdfasts (the equivalent of roots in seaweeds, which anchor the organism to the sea floor) to heat stress in the brown alga *Sargassum fusiforrme*. Under high temperature and low nutrition conditions, holdfasts can grow and regenerate into new shoots by vegetative reproduction, i.e., asexual reproduction. Thus, *S. fusiforrme* proliferates in summer via the regeneration of shoots, suggesting a relationship between high temperature tolerance and transition to the asexual growth phase. In addition, regeneration was enhanced by the fragmentation of holdfasts. Thus, the combined effects of high temperature and nutrition starvation on regeneration could be strengthened by wounding stress. The authors also demonstrate that heat stress tolerance is associated with nitrogen accumulation.

Sato et al. [27] report differences in temperature dependency of growth and sporulation in several strains of the green alga *Ulva prolifera*. Although asexual spore release was generally accelerated at 20 °C in this species, one strain did not sporulate at 20 °C, which is a notable characteristic for mariculture of *U. prolifera* in the face of increases in seawater temperature due to global warming. In addition, although heat stress generally increases nitrogen contents in this species, this strain did not show heat-stress-dependent nitrogen accumulation. Hiraoka [28] support these findings in their study comparing attached-type *U. prolifera* subsp. *prolifera* and bloom-type *Ulva prolifera* subsp. *qingdaoensis*. Although the former produces spores in spring, the latter is fragmented in spring and grows vegetatively in summer, suggesting that sporulation is inhibited under heat and nutrient starvation conditions in the bloom type. These findings indicate that heat stress tolerance is negatively related to the asexual life-cycle trade-off in *U. prolifera* subsp. *qingdaoensis*, which is in contrast to the red alga *'Bangia'* sp. ESS1, although the heat-stress-dependent promotion of vegetative growth in *U. prolifera* subsp. *qingdaoensis* is similar to that in *S. fusiforrme*. These findings again demonstrate that the relationship between stress tolerance and life-cycle trade-off differs among seaweed phyla.

Overall, the studies in this Special Issue increase our understanding of the effects of combined stresses on stress tolerance and life-cycle trade-off, which differ among species, genera, and phyla, and will contribute to the expansion and development of biological research on seaweeds. Elucidation of the mechanisms regulating stress-dependent reproductive responses will enhance our understanding of the flexible life-cycle strategies that enable seaweeds to survive in fluctuating environmental conditions by promoting the life-cycle trade-off.

Funding: This research received no external funding.

Acknowledgments: I would like to thank all authors who submitted their work for this Special Issue, as well as the reviewers for their invaluable help with manuscript evaluation and the professional editorial staff at *Plants* for their support during the completion of this issue.

Conflicts of Interest: The author declares no conflict of interest.

References

1. Wang, W.-J.; Zhu, J.-Y.; Xu, P.; Xu, J.-R.; Lin, X.-Z.; Huang, C.-K.; Song, W.-L.; Peng, G.; Wang, G.-C. Characterization of the life history of *Bangia fuscopurpurea* (Bangiaceae, Rhodophyta) in connection with its cultivation in China. *Aquaculture* **2008**, *278*, 101–109. [CrossRef]
2. Karsten, U.; West, J.A. Living in the intertidal zone seasonal effects on heterosides and sunscreen compounds in the red alga *Bangia atropurpurea* (Bangiales). *J. Exp. Mar. Bio. Ecol.* **2000**, *254*, 221–234. [CrossRef]
3. Zou, D.; Gao, K. Effects of desiccation and CO_2 concentrations on emersed photosynthesis in *Porphyra haitanensis* (Bangiales, Rhodophyta), a species farmed in China. *Eur. J. Phycol.* **2002**, *37*, 587–592. [CrossRef]
4. Thompson, R.C.; Norton, T.A.; Hawkins, S.J. Physical stress and biological control regulate the producer consumer balance in intertidal biofilms. *Ecology* **2004**, *85*, 1372–1382. [CrossRef]
5. Rawlings, T.A. Adaptations to physical stresses in the intertidal zone: The egg capsules of neogastropod molluscs. *Am. Zool.* **1999**, *39*, 230–243. [CrossRef]
6. Helmuth, B.S.; Hofmann, G.E. Microhabitats, thermal heterogeneity, and patterns of physiological stress in the rocky intertidal zone. *Biol. Bull.* **2001**, *201*, 374–384. [CrossRef]
7. Echersley, L.K.; Scrosati, R.A. Temperature, desiccation, and species performance trends along an intertidal elevation gradient. *Curr. Dev. Oceanogr.* **2012**, *5*, 59–73. [CrossRef]
8. Nakashima, K.; Yamaguchi-Shinozaki, K.; Shinozaki, K. The transcriptional regulatory network in the drought response and its crosstalk in abiotic stress responses including drought, cold, and heat. *Front. Plant Sci.* **2014**, *5*, 170. [CrossRef]
9. Rejeb, I.B.; Pastor, V.; Mauch-Mani, B. Plant responses to simultaneous biotic and abiotic stress: Molecular mechanisms. *Plants* **2014**, *3*, 458–475. [CrossRef]
10. Zhu, J.-K. Abiotic stress signaling and responses in plants. *Cell* **2016**, *167*, 313–324. [CrossRef]
11. Sade, N.; Rubio-Wilhelmi, M.D.M.; Umnajkitikorn, K.; Blumwald, E. Stress-induced senescence and plant tolerance to abiotic stress. *J. Exp. Bot.* **2018**, *69*, 845–853. [CrossRef]
12. Vishwakarma, K.; Upadhyay, N.; Kumar, N.; Yadav, G.; Singh, J.; Mishra, R.K.; Kumar, V.; Verma, R.; Upadhyay, R.G.; Pandey, M.; et al. Abscisic acid signaling and abiotic stress tolerance in plants: A review on current knowledge and future prospects. *Front. Plant Sci.* **2017**, *8*, 161. [CrossRef] [PubMed]
13. Liu, X.; Bogaert, K.; Engelen, A.H.; Leliaert, F.; Roleda, M.Y.; De Clerck, O. Seaweed reproductive biology: Environmental and genetic controls. *Bot. Mar.* **2017**, *60*, 89–108. [CrossRef]
14. Mohring, M.B.; Kendrick, G.A.; Wernberg, T.; Rule, M.J.; Vanderklift, M.A. Environmental influences on kelp performance across the reproductive period: An ecological trade-off between gametophyte survival and growth? *PLoS ONE* **2013**, *8*, e65310. [CrossRef]
15. Karasov, T.L.; Chae, E.; Herman, J.J.; Bergelson, J. Mechanisms to mitigate the trade-off between growth and defense. *Plant Cell* **2017**, *29*, 666–680. [CrossRef] [PubMed]
16. Shaar-Moshe, L.; Hayouka, R.; Roessner, U.; Peleg, Z. Phenotypic and metabolic plasticity shapes life-history strategies under combinations of abiotic stresses. *Plant Direct* **2019**, *3*, e00113. [CrossRef]
17. Barnabás, B.; Jäger, K.; Fehér, A. The effect of drought and heat stress on reproductive processes in cereals. *Plant Cell Environ.* **2008**, *31*, 11–38. [CrossRef]
18. Zinn, K.E.; Tunc-Ozdemir, M.; Harper, J.F. Temperature stress and plant sexual reproduction: Uncovering the weakest links. *J. Exp. Bot.* **2010**, *61*, 1959–1968. [CrossRef] [PubMed]
19. Hatfield, J.L.; Prueger, J.H. Temperature extremes: Effect on plant growth and development. *Weather Clim. Extrem.* **2010**, *10*, 4–10. [CrossRef]
20. Kakinuma, M.; Kaneko, I.; Coury, D.A.; Suzuki, T.; Amano, H. Isolation and identification of gametogenesis-related genes in *Porphyra yezoensis* (Rhodophyta) using subtracted cDNA libraries. *J. Appl. Phycol.* **2006**, *18*, 489–496. [CrossRef]
21. Kirk, D.L.; Kirk, M.M. Heat shock elicits production of sexual inducer in *Volvox*. *Science* **1986**, *231*, 51–54. [CrossRef] [PubMed]
22. Mikami, K.; Kishimoto, I. Temperature promoting the asexual life cycle program in *Bangia fuscopurpurea* (Bangiales, Rhodophyta) from Esashi in the Hokkaido Island, Japan. *Algal Resour.* **2018**, *11*, 25–32. [CrossRef]
23. Kishimoto, I.; Ariga, I.; Itabashi, Y.; Mikami, K. Heat-stress memory is responsible for acquired thermotolerance in *Bangia fuscopurpurea*. *J. Phycol.* **2019**, *55*, 971–975. [CrossRef] [PubMed]

24. Suda, M.; Mikami, K. Reproductive responses to wounding and heat stress in gametophytic thalli of the red alga *Pyropia yezoensis*. *Front. Mar. Sci.* **2020**, *7*, 394. [CrossRef]
25. Omuro, Y.; Khoa, H.V.; Mikami, K. The Absence of hydrodynamic stress promotes acquisition of freezing tolerance and freeze-dependent asexual reproduction in the red alga '*Bangia*' sp. ESS1. *Plants* **2021**, *10*, 465. [CrossRef]
26. Endo, H.; Sugie, T.; Yonemori, Y.; Nishikido, Y.; Moriyama, H.; Ito, R.; Okunishi, S. Vegetative reproduction is more advantageous than sexual reproduction in a canopy-forming clonal macroalga under ocean warming accompanied by oligotrophication and intensive herbivory. *Plants* **2021**, *10*, 1522. [CrossRef]
27. Sato, Y.; Kinoshita, Y.; Mogamiya, M.; Inomata, E.; Hoshino, M.; Hiraoka, M. Different growth and sporulation responses to temperature gradient among obligate apomictic strains of *Ulva Prolifera*. *Plants* **2021**, *10*, 2256. [CrossRef]
28. Hiraoka, M. Massive *Ulva* green tides caused by inhibition of biomass allocation to sporulation. *Plants* **2021**, *10*, 2482. [CrossRef]
29. Khoa, H.V.; Kumari, P.; Uchida, H.; Murakami, A.; Shimada, S.; Mikami, K. Heat-stress responses differ among species from different '*Bangia*' clades of Bangiales (Rhodophyta). *Plants* **2021**, *10*, 1733. [CrossRef]
30. Sutherland, J.; Lindstrom, S.; Nelson, W.; Brodie, J.; Lynch, M.; Hwang, M.S.; Choi, H.G.; Miyata, M.; Kikuchi, N.; Oliveira, M.; et al. A new look at an ancient order: Generic revision of the Bangiales. *J. Phycol.* **2011**, *47*, 1131–1151. [CrossRef]
31. Li, C.; Irie, R.; Shimada, S.; Mikami, K. Requirement of different normalization genes for quantitative gene expression analysis under abiotic stress conditions in '*Bangia*' sp. ESS1. *J. Aquat. Res. Mar. Sci.* **2019**, *2019*, 194–205.
32. Yokono, M.; Uchida, H.; Suzawa, Y.; Akiomoto, S.; Murakami, A. Stabilization and modulation of the phycobilisome by calcium in the calciphilic freshwater red alga *Bangia atropurpurea*. *Biochim. Biophys. Acta* **2012**, *1817*, 306–3011. [CrossRef] [PubMed]

Communication

The Absence of Hydrodynamic Stress Promotes Acquisition of Freezing Tolerance and Freeze-Dependent Asexual Reproduction in the Red Alga '*Bangia*' sp. ESS1

Yoshiki Omuro [1,†], Ho Viet Khoa [2] and Koji Mikami [3,*]

[1] Department of Aquaculture Life Science, School of Fisheries Sciences, Hokkaido University, 3-1-1 Minato-cho, Hakodate 041-8611, Japan; rattpa68@eis.hokudai.ac.jp
[2] Division of Marine Life Science, Graduate School of Fisheries Sciences, Hokkaido University, 3-1-1 Minato-cho, Hakodate 041-8611, Japan; hvkhoa59@gmail.com
[3] Department of Food Recourse Development, School of Food Industrial Sciences, Miyagi University, 2-2-1 Hatatate, Taihaku-ku, Sendai 982-0215, Japan
* Correspondence: mikamik@myu.ac.jp
† Present address: Graduate School of Science, Hokkaido University, Kita-10 Nishi-8, Kita-ku, Sapporo 060-0810, Japan.

Abstract: The ebb tide causes calm stress to intertidal seaweeds in tide pools; however, little is known about their physiological responses to loss of water movement. This study investigated the effects of static culture of '*Bangia*' sp. ESS1 at 15 °C on tolerance to temperature fluctuation. The freezing of aer-obically cultured thalli at −80 °C for 10 min resulted in the death of most cells. By contrast, statically cultured thalli acquired freezing tolerance that increased cell viability after freeze–thaw cycles, although they did not achieve thermotolerance that would enable survival at the lethal temperature of 32 °C. Consistently, the unsaturation of membrane fatty acids occurred in static culture. Notably, static culture of thalli enhanced the release of asexual spores after freeze-and-thaw treatment. We conclude that calm stress triggers both the acquisition of freezing tolerance and the promotion of freezing-dependent asexual reproduction. These findings provide novel insights into stress toler-ance and the regulation of asexual reproduction in Bangiales.

Keywords: asexual reproduction; '*Bangia*' sp. ESS1; Bangiales; calm stress; freezing tolerance; fatty acid; membrane fluidity

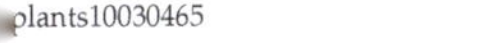

Citation: Omuro, Y.; Khoa, H.V.; Mikami, K. The Absence of Hydrodynamic Stress Promotes Acquisition of Freezing Tolerance and Freeze-Dependent Asexual Reproduction in the Red Alga '*Bangia*' sp. ESS1. *Plants* **2021**, *10*, 465. https://doi.org/10.3390/plants10030465

Academic Editor: Dariusz Latowski

Received: 5 February 2021
Accepted: 26 February 2021
Published: 1 March 2021

Publisher's Note: MDPI stays neutral with regard to jurisdictional claims in published maps and institutional affiliations.

Copyright: © 2021 by the authors. Licensee MDPI, Basel, Switzerland. This article is an open access article distributed under the terms and conditions of the Creative Commons Attribution (CC BY) license (https://creativecommons.org/licenses/by/4.0/).

1. Introduction

Bangiales is an order of red algae characterized by thalli with a filamentous or leafy shape [1,2]. These seaweeds are sessile multicellular organisms that live in intertidal regions, where temperature, salinity, and nutritional conditions usually fluctuate. Recent physiological and "omics" analyses indicate that Bangiales respond to heat, cold, salinity, hyper-osmolality, and desiccation through stress-inducible gene expression and repression [3–9]. Thus, Bangiales sense environmental changes as different abiotic stresses and express or repress different sets of genes for each stress. Since thalli of Bangiales mostly appear in winter and early spring, acclimation to low-temperature stress seems to be essential for their growth and survival; however, little is known about how Bangiales acquire tolerance to cold stress.

Cold acclimation is a phenomenon in which the exposure of plants to non-freezing (chilling) temperature promotes the acquisition of tolerance to freezing at sub-zero temperature [10–14]. Cold acclimation has been observed in both micro- and macroalgae [15–17]. In terrestrial plants, cold acclimation is established via exposure to other stresses [18–20]. For instance, desiccation stress induced freezing tolerance in winter cereals [21–23]. A similar phenomenon has been observed in microalgae [17]. These findings suggest that the ability to acquire freezing tolerance by exposure to environmental stresses other than low temperature might be conserved among photosynthetic organisms. However, it has not yet been examined whether cold acclimation is established by such a cross-tolerance mechanism in macroalgae.

There is a close relationship between an increase in cold-stress tolerance and membrane fluidization via the unsaturation of membrane fatty acids in poikilothermic organisms [24–26], which can be demonstrated by the artificial unsaturation of membrane fatty acids via genetic transformation using fatty acid desaturase genes [27–34]. Recently, cold stress-induced unsaturation of membrane fatty acids was also reported in the red seaweed *Bangia fuscopurpurea* [5]. Thus, the monitoring of changes in the membrane fatty acid composition can serve as a powerful tool to evaluate physiological responses related to freezing tolerance via cold acclimation in algae.

The ebb tide and resulting loss of water flow—the most drastic change in living conditions at the intertidal region—exposes Bangiales to temperature changes, desiccation, nutritional starvation, and other potential stresses. We hypothesized that loss of water movement might strengthen the effects of environmental changes on growth and viability and thus trigger the acquisition of tolerance to abiotic stresses in Bangiales. The filamentous red seaweed '*Bangia*' sp. ESS1 [35] is used as a model organism to investigate the stress responses of Bangiales in our laboratory. Using this species, we previously reported an acceleration of asexual reproduction under heat-stress conditions [36] and confirmed that heat-stress memory has an intrinsic ability to induce thermotolerance [37]. We also established a transient gene expression system [38] and identified reference genes to quantify gene expression under various kinds of abiotic stress [35]. Therefore, to test our hypothesis, we employed '*Bangia*' sp. ESS1 and focused on loss of water movement due to ebb tide as an abiotic stress. We investigated the effects of static culture at 15 °C, a regular laboratory culture temperature, on the acquisition of tolerance to temperature fluctuation and on membrane fatty acid compositions.

2. Results

2.1. Acquisition of Freezing Tolerance by Exposure to Calm Stress

When thalli of '*Bangia*' sp. ESS1 were grown under hydrodynamic stress by aeration culture at 15 °C, frozen at −80 °C for 10 min, and then returned to 15 °C seawater, most of the cells died (25% viability as shown in Figure 1A). Thus, aeration-cultured '*Bangia*' sp. ESS1 has little tolerance to direct transfer from 15 °C to sub-zero temperature. By contrast, when aeration-cultured thalli were statically cultured at 15 °C for 1–6 weeks prior to freezing, cell viability was significantly higher, gradually increasing with the duration of static culture to 90% (Figure 1A). Viability was maintained for 1 week after freeze-and-thaw treatment (Figure S1). In addition, when similarly treated samples were directly transferred to 32 °C seawater, as a lethal heat-stress condition [36], after freezing, viability decreased depending on the duration of heat-stress exposure (Figure S2). Moreover, static culture of thalli at 15 °C for 2 weeks or 6 weeks accelerated the release of asexual spores after freeze-and-thaw treatment and subsequent 1 week-culture at 15 °C; this did not occur in statically cultured thalli without freezing (Figure 1B).

2.2. Unsaturation of Membrane Fatty Acids under Calm Stress Conditions

To examine whether the membrane fatty acid composition is modulated by calm conditions, we incubated aeration-cultured thalli at 15 °C in static culture for 1–6 weeks and analyzed membrane fatty acid compositions of samples harvested at every week. The relative amounts of saturated fatty acids and monoenes decreased compared to those in aeration-cultured thalli, whereas the relative amounts of polyenes gradually increased (Figure S3). The significance of decreases in saturated fatty acids and monoenes and increases in polyenes was clearly demonstrated by comparison of the fatty acid compositions among aeration-cultured thalli and samples that were statically cultured for 2 weeks or 6 weeks (Figure 2). Since the main saturated and unsaturated fatty acids were palmitic acid (16:0) and eicosapentaenoic acid (20:5 *n*-3), respectively (data not shown), the results of saturates and polyenes in Figure 2 roughly reflected the changes in contents of these fatty acids.

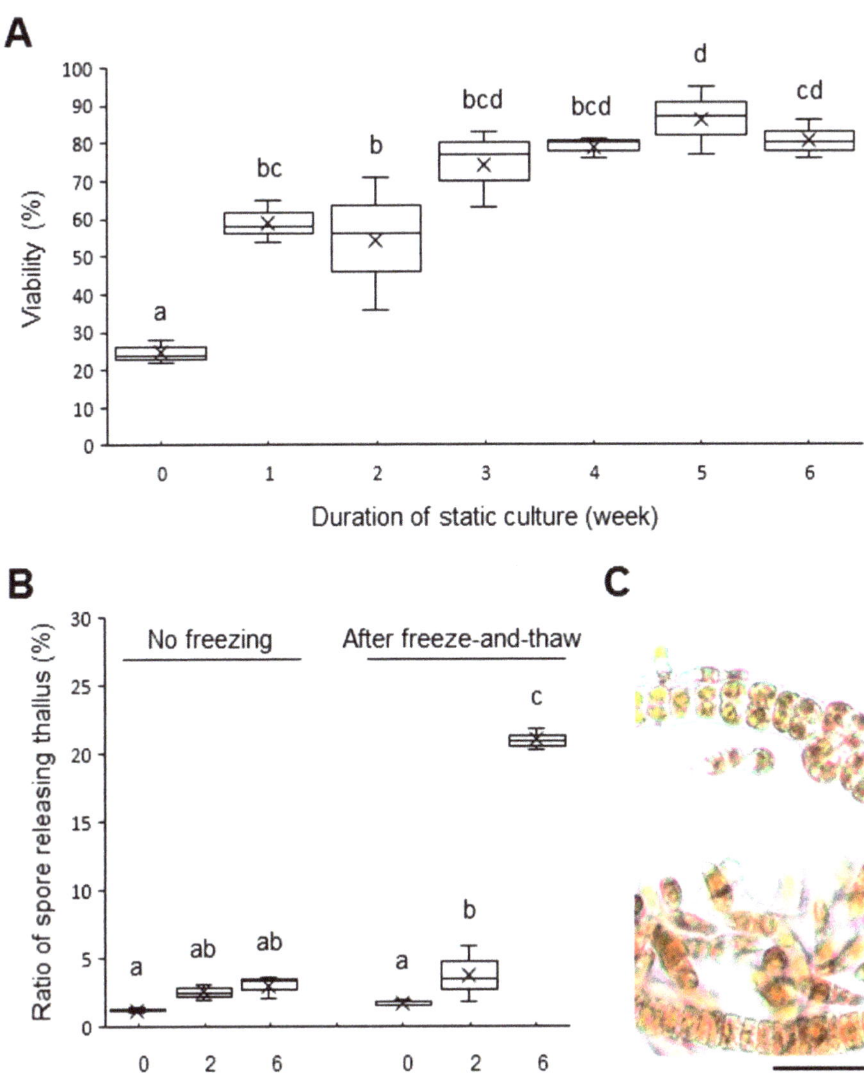

Figure 1. Effects of static culture on viability and release of asexual spores in *'Bangia'* sp. ESS1. (**A**) Static culture-induced increase in viability after freeze-and-thaw. (**B**) Enhancement of release of asexual spores from statically cultured thalli after freeze-and-thaw. (**C**) Extensive release of asexual spores in thalli exposed to freeze-and-thaw treatment (lower) than non-frozen thalli (upper) after static culture for 6 weeks. Most released spores developed into small germlings. Scale bar: 50 μm. Letters on boxes denote significant differences from triplicate experiments defined by the Tukey–Kramer test ($p < 0.05$) in one-way ANOVA.

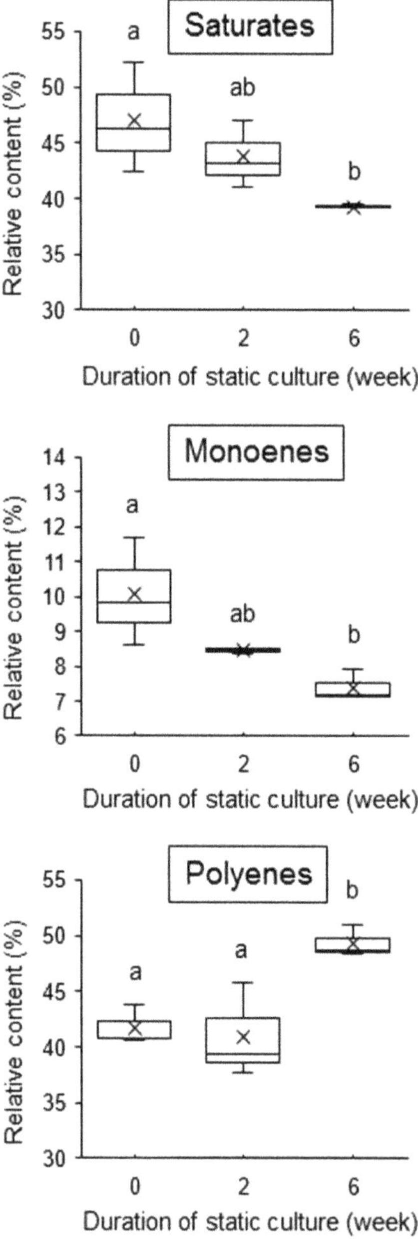

Figure 2. Effects of static culture on membrane fatty acid compositions in *'Bangia'* sp. ESS1. Changes in the relative amounts of saturated fatty acids (Saturates), monounsaturated fatty acids (Monoenes), and polyunsaturated fatty acids (Polyenes) were analyzed in thalli statically cultured for 2 and 6 weeks in comparison with control samples (0) without static culture. Letters on boxes denote significant differences from triplicate experiments defined by the Tukey–Kramer test ($p < 0.05$) in one-way ANOVA.

3. Discussion

We here demonstrated that calm stress promotes the acquisition of freezing tolerance and an increase in the unsaturation of membrane fatty acids, which enables survival upon exposure to −80 °C in *'Bangia'* sp. ESS1 but does not induce thermotolerance. Remarkably, freezing tolerance was established with only 1 week of static culture. Thus, loss of water movement can mimic chilling as a priming stress that triggers the establishment of freezing tolerance, meaning that the acquisition of freezing tolerance is one strategy for the toleration of calm conditions. This finding is consistent with our previous results showing that an increased saturation of membrane fatty acids is required for the acquisition of heat-stress tolerance [37], which is reciprocally related to the decrease in saturation level by static culture shown in Figure 2 and suggests that membrane fluidization is critically involved in calm-stress responses. Although the positive contribution of calm conditions to gamete release has been demonstrated in green and brown algae [39–41], the finding that calm stress promotes the acquisition of freezing tolerance in algae is novel.

Our results also indicate that freeze-and-thaw treatment of statically cultured gametophytes enhances the release of asexual spores in *'Bangia'* sp. ESS1, in which asexual reproduction is accelerated in a freezing-dependent manner. Thus, loss of water movement seems to increase sensitivity to freeze-and-thaw cycles for promotion of the asexual life cycle. We previously observed an enhancement of asexual reproduction by heat stress in this alga [36] and by hypo-osmotic, oxidative, and wounding stresses in the red alga *Pyropia yezoensis* [42–44]. Thus, we propose that environmental stress can trigger a transition from growth to reproductive phase in the life cycle of Bangiales. Consistently, gamete release by the depletion of dissolved inorganic carbon (DIC) was previously proposed under calm conditions in the brown alga *Fucus distichus* [39]; however, it is unknown whether reduced DIC content acts as a signal to promote the acquisition of freezing tolerance in *'Bangia'* sp. ESS1. Therefore, studying the regulatory mechanisms underlying spore formation and release will further elucidate abiotic stress-inducible asexual reproduction and its relation to membrane fluidity in Bangiales.

Since hydrodynamic stress essentially occurs in the hydrosphere, an ability to acquire cross-tolerance to calm and freezing stresses seems to be unique to aquatic organisms. The acquisition of freezing tolerance by exposure to calm stress is a reasonable adaptation to the circumstances of a tide pool, where organisms may experience falling air temperatures and snow. However, little is known about the sensing of and signal transduction in response to loss of water movement in algae. Thus, the identification of factors that trigger and/or participate in freezing tolerance and membrane fatty acid unsaturation in response to loss of water movement could help us understand how calm-stress signaling is regulated and interacts with chilling signal transduction pathways in *'Bangia'* sp. ESS1.

4. Materials and Methods

4.1. Algal Material and Stress Treatment

Thalli of the marine red seaweed *'Bangia'* sp. ESS1 [35] were collected at Esashi, Hokkaido, Japan on 14 May 2010 [38], and a clean single thallus of unknown sex was aeration cultured at 15 °C and maintained as an experimental line. For static culture, 0.1 g (fresh weight) samples of thalli were cultured in dishes (Petri dish φ90 × 20 mm height) containing 50 mL of enriched SEALIFE (ESL) medium [35] at 15 °C under 60 μmol m^{-2} s^{-1} irradiation for 0 (control) to 6 weeks. Algal samples harvested every week were stored at −80 °C for 10 min and then cultured at 15 °C for 0 (just after thawing) and 1 week as a freezing-stress treatment or cultured at a lethal temperature of 32 °C for 0, 1, 3, 5, and 7 days as a heat-stress treatment. Samples harvested at every week of freezing treatment and all durations of heat-stress treatment were subjected to analyses of viability and membrane fatty acid composition as described below.

4.2. Viability Test and Observation of Asexual Spore Release

The viability of cells of statically cultured and freeze-and-thaw treated thalli was examined as described previously [36,37]. In brief, '*Bangia*' sp. ESS1 thalli exposed to calm and freezing stresses as described above were stained daily with ESL medium containing 0.01% erythrosine (Wako Pure Chemical Industries, Osaka, Japan). After staining for 20 min at room temperature, thalli were gently rinsed with ESL medium to remove excess erythrosine and mounted on slides with ESL medium. Thalli were observed and photographed using an Olympus IX73 light microscope equipped with an Olympus DP22 camera. Cells stained by the dye were defined as dead cells, as indicated in Takahashi et al. [45]. Viability was calculated from the number of living and dead cells obtained using micrographs. The observation of asexual spore release from statically cultured and freeze-and-thaw treated thalli was performed microscopically as described above. The ratio of the number of asexual spore-releasing thalli to the total number of thalli was calculated.

4.3. Analysis of Membrane Fatty Acid Composition

Fresh samples of '*Bangia*' sp. ESS1 were immersed in boiling water for 3 min to deactivate lipid hydrolytic enzymes and then freeze-dried and homogenized using a grinder. Lipids were extracted from 0.1 g powdered algal sample via the Bligh–Dyer method [46] with some modifications as described in Kishimoto et al. [37]. The preparation of fatty acid methyl esters based on Christie and Han [47] and its GC analysis using a Shimadzu GC-14A gas chromatograph (Shimadzu Corporation, Kyoto, Japan) with an Omegawax 320 column (30 m × 0.32 mm i.d., Supeleo, PA, USA) were performed as described previously [37].

4.4. Statistical Analysis

Values are indicated with SD from triplicate experiments. A one-way ANOVA followed by a Tukey–Kramer test was used for multiple comparisons, and significant differences were determined using a cutoff value of $p < 0.05$ as described in [48].

5. Conclusions

'*Bangia*' sp. ESS1 acquires freezing tolerance when it is exposed to calm stress, for which an increase in unsaturation levels of membrane fatty acids might be involved. Recent studies have indicated the involvement of membrane integrity based on lipid remodeling in freezing tolerance in terrestrial plants [26,49–51], although changes in membrane lipid compositions by transferring from aeration to static culture conditions were not analyzed in algae. Therefore, elucidation of the relationship between calm-stress signaling and lipid remodeling in membranes under null hydrodynamic stress conditions in '*Bangia*' sp. ESS1 could provide insights into the unique characteristics that regulate the acquisition of freezing tolerance and asexual life cycle by calm stress in Bangiales.

Supplementary Materials: The following are available online at https://www.mdpi.com/2223-7747/10/3/465/s1, Figure S1: Maintenance of viability after freezing due to static culture in '*Bangia*' sp. ESS1, Figure S2: Effects of static culture on viability under lethal heat conditions in '*Bangia*' sp. ESS1, Figure S3: Gradual changes in membrane fatty acid composition by static culture of '*Bangia*' sp. ESS1.

Author Contributions: Conceptualization, K.M.; methodology, K.M. and Y.O.; validation, Y.O and H.V.K.; formal analysis, Y.O and H.V.K.; investigation, Y.O and H.V.K.; data curation, Y.O. and H.V.K.; writing—original draft preparation and writing—review and editing, K.M.; visualization, Y.O., H.V.K. and K.M.; supervision and project administration, K.M. All authors have read and agreed to the published version of the manuscript.

Funding: This research received no external funding.

Data Availability Statement: Data is contained within the article.

Acknowledgments: We thank Puja Kumari for technical supports on fatty acid analysis.

Conflicts of Interest: The authors declare no conflict of interest.

References

1. Sutherland, J.E.; Lindstrom, S.C.; Nelson, W.A.; Brodie, J.; Lynch, M.D.J.; Hwang, M.S.; Choi, H.-G.; Miyata, M.; Kikuchi, N.; Oliveira, M.C.; et al. A new look at an ancient order: Generic revision of the Bangiales (Rhodophyta). *J. Phycol.* **2011**, *47*, 1131–1151. [CrossRef]
2. Yang, L.E.; Lu, Q.Q.; Brodie, J. A review of the bladed Bangiales (Rhodophyta) in China: History, culture and taxonomy. *Eur. J. Phycol.* **2017**, *52*, 251–263. [CrossRef]
3. Sun, P.; Mao, Y.; Li, G.; Cao, M.; Kong, F.; Wang, L.; Bi, G. Comparative transcriptome profiling of *Pyropia yezoensis* (Ueda) M.S. Hwang & H.G. Choi in response to temperature stresses. *BMC Genom.* **2015**, *16*, 463.
4. Sun, P.; Tang, X.; Bi, G.; Xu, K.; Kong, F.; Mao, Y. Gene expression profiles of *Pyropia yezoensis* in response to dehydration and rehydration stresses. *Mar. Genomics* **2019**, *43*, 43–49. [CrossRef]
5. Cao, M.; Wang, D.; Mao, Y.; Kong, F.; Bi, G.; Xing, Q.; Weng, Z. Integrating transcriptomics and metabolomics to characterize the regulation of EPA biosynthesis in response to cold stress in seaweed *'Bangia' fuscopurpurea*. *PLoS ONE* **2017**, *12*, e0186986. [CrossRef] [PubMed]
6. Gao, D.; Kong, F.; Sun, P.; Bi, G.; Mao, Y. Transcriptome-wide identification of optimal reference genes for expression analysis of *Pyropia yezoensis* responses to abiotic stress. *BMC Genom.* **2018**, *19*, 251. [CrossRef] [PubMed]
7. Wang, W.; Teng, F.; Lin, Y.; Ji, D.; Xu, Y.; Chen, C.; Xie, C. Transcriptomic study to understand thermal adaptation in a high temperature-tolerant strain of *Pyropia haitanensis*. *PLoS ONE* **2018**, *13*, e0195842. [CrossRef]
8. Wang, W.; Shen, Z.; Sun, X.; Liu, F.; Liang, Z.; Wang, F.; Zhu, J. De novo transcriptomics analysis revealed a global reprogramming towards dehydration and hyposalinity in *'Bangia' fuscopurpurea* gametophytes (Rhodophyta). *J. Appl. Phycol.* **2019**, *31*, 637–651. [CrossRef]
9. Wang, W.; Xu, Y.; Chen, T.; Xing, L.; Xu, K.; Xu, Y.; Ji, D.; Chen, C.; Xie, C. Regulatory mechanisms underlying the maintenance of homeostasis in *Pyropia haitanensis* under hypersaline stress conditions. *Sci. Total Environ.* **2019**, *662*, 168–179. [CrossRef] [PubMed]
10. Wanner, L.A.; Junttila, O. Cold-induced freezing tolerance in Arabidopsis. *Plant Physiol.* **1999**, *120*, 391–400. [CrossRef] [PubMed]
11. Kaplan, F.; Kopka, J.; Haskell, D.W.; Zhao, W.; Schiller, K.C.; Gatzke, N.; Sung, D.Y.; Guy, C.L. Exploring the temperature-stress metabolome of Arabidopsis. *Plant Physiol.* **2004**, *136*, 4159–4168. [CrossRef]
12. Tan, T.; Sun, Y.; Peng, X.; Wu, G.; Bao, G.; He, F.; Zhou, H.; Lin, H. *ABSCISIC ACID INSENSITIVE3* is involved in cold response and freezing tolerance regulation in *Physcomitrella patens*. *Front. Plant Sci.* **2017**, *8*, 1599.
13. Thomashow, M.F. PLANT COLD ACCLIMATION: Freezing tolerance genes and regulatory mechanisms. *Annu. Rev. Plant Physiol. Plant Mol. Biol.* **1999**, *50*, 571–599. [CrossRef] [PubMed]
14. Miura, K.; Furumoto, T. Cold signaling and cold response in plants. *Int. J. Mol. Sci.* **2013**, *14*, 5312–5337. [CrossRef] [PubMed]
15. Ben-Amotz, A.; Gilboa, A. Cryopreservation of marine unicellular algae. II. Induction of freezing tolerance. *Mar. Ecol.* **1980**, *2*, 157161. [CrossRef]
16. Nagao, M.; Matsui, K.; Uemura, M. *Klebsormidium flaccidum*, a charophycean green alga, exhibits cold acclimation that is closely associated with compatible solute accumulation and ultrastructural changes. *Plant Cell Environ.* **2008**, *31*, 872–885. [CrossRef]
17. Anandarajah, K.; Perumal, G.M.; Sommerfeld, M.; Hu, Q. Induced freezing and desiccation tolerance in the microalgae wild type *Nannochloropsis* sp. and *Scenedesmus dimorphus*. *Aust. J. Basic Appl. Sci.* **2011**, *5*, 678–686.
18. Hossain, M.A.; Bhattacharjee, S.; Armin, S.M.; Qian, P.; Xin, W.; Li, H.Y.; Burritt, D.J.; Fujita, M.; Tran, L.S. Hydrogen peroxide priming modulates abiotic oxidative stress tolerance: Insights from ROS detoxification and scavenging. *Front. Plant Sci.* **2015**, *6*, 420. [CrossRef]
19. Foyer, C.H.; Rasool, B.; Davey, J.W.; Hancock, R.D. Cross-tolerance to biotic and abiotic stresses in plants: A focus on resistance to aphid infestation. *J. Exp. Bot.* **2016**, *67*, 2025–2037. [CrossRef]
20. Yasuda, H. Cross-tolerance to thermal stresses and its application to the development of cold tolerant rice. *Jpn. Agric. Res. Q.* **2017**, *51*, 99–105. [CrossRef]
21. Siminovitch, D.; Cloutier, Y. Twenty-four-hour induction of freezing and drought tolerance in plumules of winter rye seedlings by desiccation stress at room temperature in the dark. *Plant Physiol.* **1982**, *69*, 250–255. [CrossRef] [PubMed]
22. Cloutier, Y.; Andrews, C.J. Efficiency of cold hardiness induction by desiccation stress in four winter cereals. *Plant Physiol.* **1984**, *76*, 595–598. [CrossRef] [PubMed]
23. Hussain, H.A.; Hussain, S.; Khaliq, A.; Ashraf, U.; Anjum, S.A.; Men, S.; Wang, L. Chilling and drought stresses in crop plants: Implications, cross talk, and potential management opportunities. *Front. Plant Sci.* **2018**, *9*, 393. [CrossRef]
24. Mikami, K.; Murata, N. Membrane fluidity and the perception of environmental signals in cyanobacteria and plants. *Prog. Lipid. Res.* **2003**, *42*, 527–543. [CrossRef]
25. Guschina, I.A.; Harwood, J.L. Lipids and lipid metabolism in eukaryotic algae. *Prog. Lipid. Res.* **2006**, *45*, 160–186. [CrossRef]
26. Barrero-Sicilia, C.; Silvestre, S.; Haslam, R.P.; Michaelson, L.V. Lipid remodelling: Unravelling the response to cold stress in Arabidopsis and its extremophile relative *Eutrema salsugineum*. *Plant Sci.* **2017**, *263*, 194–200. [CrossRef]
27. Kodama, H.; Hamada, T.; Horiguchi, G.; Nishimura, M.; Iba, K. Genetic enhancement of cold tolerance by expression of a gene for chloroplast [omega]-3 fatty acid desaturase in transgenic tobacco. *Plant Physiol.* **1994**, *105*, 601–605. [CrossRef] [PubMed]
28. Zhang, M.; Barg, R.; Yin, M.; Gueta-Dahan, Y.; Leikin-Frenkel, A.; Salts, Y.; Shabtai, S.; Ben-Hayyim, G. Modulated fatty acid desaturation via overexpression of two distinct omega-3 desaturases differentially alters tolerance to various abiotic stresses in transgenic tobacco cells and plants. *Plant J.* **2005**, *44*, 361–371. [CrossRef]

29. Khodakovskaya, M.; McAvoy, R.; Peters, J.; Wu, H.; Li, Y. Enhanced cold tolerance in transgenic tobacco expressing a chloroplast omega-3 fatty acid desaturase gene under the control of a cold-inducible promoter. *Planta* **2006**, *223*, 1090–1100. [CrossRef]
30. Yu, C.; Wang, H.S.; Yang, S.; Tang, X.F.; Duan, M.; Meng, Q.W. Overexpression of endoplasmic reticulum omega-3 fatty acid desaturase gene improves chilling tolerance in tomato. *Plant Physiol. Biochem.* **2009**, *47*, 1102–1112. [CrossRef]
31. Román, Á.; Andreu, V.; Hernández, M.L.; Lagunas, B.; Picorel, R.; Martínez-Rivas, J.M.; Alfonso, M. Contribution of the different omega-3 fatty acid desaturase genes to the cold response in soybean. *J. Exp. Bot.* **2012**, *63*, 4973–4982. [CrossRef]
32. Shi, J.; Cao, Y.; Fan, X.; Li, M.; Wang, Y.; Ming, F. A rice microsomal delta-12 fatty acid desaturase can enhance resistance to cold stress in yeast and *Oryza sativa*. *Mol. Breed.* **2012**, *29*, 743–757. [CrossRef]
33. Shi, Y.; Yue, X.; An, L. Integrated regulation triggered by a cryophyte ω-3 desaturase gene confers multiple-stress tolerance in tobacco. *J. Exp. Bot.* **2018**, *69*, 2131–2148. [CrossRef]
34. Xue, M.; Guo, T.; Ren, M.; Wang, Z.; Tang, K.; Zhang, W. Constitutive expression of chloroplast glycerol-3-phosphate acyltransferase from *Ammopiptanthus mongolicus* enhances unsaturation of chloroplast lipids and tolerance to chilling, freezing and oxidative stress in transgenic *Arabidopsis*. *Plant Physiol. Biochem.* **2019**, *143*, 375–387. [CrossRef]
35. Li, C.; Irie, R.; Shimada, S.; Mikami, K. Requirement of different normalization genes for quantitative gene expression analysis under abiotic stress conditions in '*Bangia*' sp. ESS1. *J. Aquat. Res. Mar. Sci.* **2019**, *2019*, 194–205.
36. Mikami, K.; Kishimoto, I. Temperature promoting the asexual life cycle program in '*Bangia*' *fuscopurpurea* (Bangiales, Rhodophyta) from Esashi in the Hokkaido Island, Japan. *Algal Resour.* **2018**, *11*, 25–32.
37. Kishimoto, I.; Ariga, I.; Itabashi, Y.; Mikami, K. Heat-stress memory is responsible for acquired thermotolerance in *Bangia fuscopurpurea*. *J. Phycol.* **2019**, *55*, 971–975. [CrossRef] [PubMed]
38. Hirata, R.; Takahashi, M.; Saga, N.; Mikami, K. Transient gene expression system established in *Porphyra yezoensis* is widely applicable in Bangiophycean algae. *Mar. Biotechnol.* **2011**, *13*, 1038–1047. [CrossRef] [PubMed]
39. Pearson, G.A.; Serrão, E.A.; Brawley, S.H. Control of gamete release in fucoid algae: Sensing hydrodynamic conditions via carbon acquisition. *Ecology* **1998**, *79*, 1725–1739. [CrossRef]
40. Brawley, S.H.; Johnson, L.E.; Pearson, G.A.; Speransky, V.; Li, R.; Serrão, E.A. Gamete release at low tide in fucoid algae: Maladaptive or advantageous? *Am. Zool.* **1999**, *39*, 218–229. [CrossRef]
41. Speransky, S.R.; Brawley, S.H.; Halteman, W.A. Gamete re-lease is increased by calm conditions in the coenocytic green alga *Bryopsis* (Chlorophyta). *J. Phycol.* **2000**, *36*, 730–739. [CrossRef]
42. Li, L.; Saga, N.; Mikami, K. Phosphatidylinositol 3-kinase activity and asymmetrical accumulation of F-actin are necessary for establishment of cell polarity in the early development of monospores from the marine red alga *Porphyra yezoensis*. *J. Exp. Bot.* **2008**, *59*, 3575–3586. [CrossRef]
43. Takahashi, M.; Mikami, K. Oxidative stress promotes asexual reproduction and apogamy in the red seaweed *Pyropia yezoensis*. *Front. Plant Sci.* **2017**, *8*, 62. [CrossRef]
44. Suda, M.; Mikami, K. Reproductive responses to wounding and heat stress in gametophytic thalli of the red alga *Pyropia yezoensis*. *Front. Mar. Sci.* **2020**, *7*, 394. [CrossRef]
45. Takahashi, M.; Mikami, K.; Mizuta, H.; Saga, N. Identification and efficient utilization of antibiotics for the development of a stable transformation system in *Porphyra yezoensis* (Bangia, Rhodophyta). *J. Aquac. Res. Dev.* **2011**, *2*, 118. [CrossRef]
46. Bligh, E.G.; Dyer, W.J. A rapid method of lipid extraction and purification. *Can. J. Biochem. Physiol.* **1959**, *37*, 911–917. [CrossRef] [PubMed]
47. Christie, W.W.; Han, X. Preparation of derivatives of fatty acids. In *Lipid Analysis*, 4th ed.; Christie, W.W., Han, X., Eds.; The Oily Press: Bridgwater, UK, 2010; pp. 145–158.
48. Mikami, K.; Ito, M.; Taya, K.; Kishimoto, I.; Kobayashi, T.; Itabashi, Y.; Tanaka, R. Parthenosporophytes of the brown alga *Ectocarpus siliculosus* exhibit sex-dependent differences in thermotolerance as well as fatty acid and sterol composition. *Mar. Environ. Res.* **2018**, *137*, 188–195. [CrossRef]
49. Arisz, S.A.; Heo, J.Y.; Koevoets, I.T.; Zhao, T.; van Egmond, P.; Meyer, A.J.; Zeng, W.; Niu, X.; Wang, B.; Mitchell-Olds, T.; et al. DIACYLGLYCEROL ACYLTRANSFERASE1 contributes to freezing tolerance. *Plant Physiol.* **2018**, *177*, 1410–1424. [CrossRef] [PubMed]
50. Tan, W.-J.; Yang, Y.-C.; Zhou, Y.; Huang, L.-P.; Xu, L.; Chen, Q.-F.; Yu, L.-J.; Xiao, S. DIACYLGLYCEROL ACYLTRANSFERASE and DIACYLGLYCEROL KINASE modulate triacylglycerol and phosphatidic acid production in the plant response to freezing stress. *Plant Physiol.* **2018**, *177*, 1303–1318. [CrossRef] [PubMed]
51. Zhang, H.; Dong, J.; Zhao, X.; Zhang, Y.; Ren, J.; Xing, L.; Jiang, C.; Wang, X.; Wang, J.; Zhao, S.; et al. Research progress in membrane lipid metabolism and molecular mechanism in peanut cold tolerance. *Front. Plant Sci.* **2019**, *10*, 838. [CrossRef] [PubMed]

plants

MDPI

Article

Heat-Stress Responses Differ among Species from Different 'Bangia' Clades of Bangiales (Rhodophyta)

Ho Viet Khoa [1], Puja Kumari [2,3], Hiroko Uchida [4], Akio Murakami [4,5], Satoshi Shimada [6] and Koji Mikami [7,*]

1 Graduate School of Fisheries Sciences, Hokkaido University, 3-1-1 Minato-cho, Hakodate 041-8611, Japan; hvkhoa59@gmail.com
2 Faculty of Fisheries Sciences, Hokkaido University, 3-1-1 Minato-cho, Hakodate 041-8611, Japan; pujamashal@gmail.com
3 School of Biological Sciences, University of Aberdeen, Cruickshank Building, St. Machar Drive, Aberdeen AB24 3UU, UK
4 Kobe University Research Center for Inland Seas, 2746 Iwaya, Awaji 656-2401, Japan; ucchy@maia.eonet.ne.jp (H.U.); akiomura@kobe-u.ac.jp (A.M.)
5 Graduate School of Science, Department of Biology, Kobe University, 1-1 Rokkodai-cho, Nada-ku, Kobe 657-8501, Japan
6 Department of Biology, Ochanomizu University, 2-1-1 Otsuka, Bunkyo-ku, Tokyo 112-8610, Japan; satoshimiru@gmail.com
7 Department of Integrative Studies of Plant and Animal Production, School of Food Industrial Sciences, Miyagi University, 2-2-1 Hatatate, Taihaku-ku, Sendai 982-0215, Japan
* Correspondence: mikamik@myu.ac.jp

Citation: Khoa, H.V.; Kumari, P.; Uchida, H.; Murakami, A.; Shimada, S.; Mikami, K. Heat-Stress Responses Differ among Species from Different '*Bangia*' Clades of Bangiales (Rhodophyta). *Plants* **2021**, *10*, 1733. https://doi.org/10.3390/plants10081733

Academic Editor: Antonino Pollio

Received: 5 August 2021
Accepted: 20 August 2021
Published: 22 August 2021

Publisher's Note: MDPI stays neutral with regard to jurisdictional claims in published maps and institutional affiliations.

Copyright: © 2021 by the authors. Licensee MDPI, Basel, Switzerland. This article is an open access article distributed under the terms and conditions of the Creative Commons Attribution (CC BY) license (https://creativecommons.org/licenses/by/4.0/).

Abstract: The red alga '*Bangia*' sp. ESS1, a '*Bangia*' 2 clade member, responds to heat stress via accelerated asexual reproduction and acquires thermotolerance based on heat-stress memory. However, whether these strategies are specific to '*Bangia*' 2, especially '*Bangia*' sp. ESS1, or whether they are employed by all '*Bangia*' species is currently unknown. Here, we examined the heat-stress responses of '*Bangia*' sp. ESS2, a newly identified '*Bangia*' clade 3 member, and *Bangia atropurpurea*. Intrinsic thermotolerance differed among species: Whereas '*Bangia*' sp. ESS1 survived at 30 °C for 7 days, '*Bangia*' sp. ESS2 and *B. atropurpurea* did not, with *B. atropurpurea* showing the highest heat sensitivity. Under sublethal heat stress, the release of asexual spores was highly repressed in '*Bangia*' sp. ESS2 and completely repressed in *B. atropurpurea*, whereas it was enhanced in '*Bangia*' sp. ESS1. '*Bangia*' sp. ESS2 failed to acquire heat-stress tolerance under sublethal heat-stress conditions, whereas the acquisition of heat tolerance by priming with sublethal high temperatures was observed in both *B. atropurpurea* and '*Bangia*' sp. ESS1. Finally, unlike '*Bangia*' sp. ESS1, neither '*Bangia*' sp. ESS2 nor *B. atropurpurea* acquired heat-stress memory. These findings provide insights into the diverse heat-stress response strategies among species from different clades of '*Bangia*'.

Keywords: *Bangia atropurpurea*; '*Bangia*' sp.; heat stress; asexual reproduction; stress memory; thermotolerance

1. Introduction

Bangiales is a monophyletic order of red algae [1] comprising over 150 species [2]. Although filamentous *Bangia* Lyngb. and foliose *Porphyra* C. Agardh were previously recognized as genera within Bangiales [3], recent phylogenetic analyses demonstrated the presence of unexpected diversity in Bangiales, with cryptic species showing highly similar morphologies [4–9]. This limits the taxonomic analysis of Bangiales based on morphological characteristics. Therefore, to understand the diversity within Bangiales, it is important to compare the nucleotide sequences of widely conserved genes, such as the plastid gene *rbcL*, encoding the large subunit of ribulose 1,5-bisphosphate carboxylase/oxygenase (Rubisco), and the nucleus-encoded small subunit ribosomal ribonucleic acid (SSU rRNA).

Based on phylogenetic analyses of *rbcL* and SSU rRNA [10,11], the taxonomy of Bangiales was revised by splitting it into 13 genera (including 4 filamentous and 9 foliose genera) and 3 filamentous clades. Accordingly, although for the past few decades, all members of the genus *Bangia* were considered to include only three species, i.e., *B. atropurpurea*, *B. fuscopurpurea*, and *B. gloiopeltidicola*, the previously recognized genus *Bangia* was recently divided into one genus (*Bangia*) and three clades (*'Bangia'* 1, *'Bangia'* 2, and *'Bangia'* 3) [10]. In this classification system, *B. atropurpurea* and *B. gloiopeltidicola* fall into *Bangia* and *'Bangia'* 3, respectively, while *B. fuscopurpurea*, which has been used in numerous studies and is thought to represent a single species, is classified into *'Bangia'* 2 and *'Bangia'* 3 [10]. Thus, species previously classified as *B. fuscopurpurea* are a mixture of phylogenetically close but distinct species.

In Japan, species belonging to *'Bangia'* are widely distributed along coastal regions. However, the local and seasonal distributions of these species have not been extensively surveyed. It is therefore unclear whether species belonging to the three *'Bangia'* clades are present in Japan, although the *'Bangia'* 2 species, *'Bangia'* sp. ESS1, has been identified in Esashi, Hokkaido [12]. The use of a combination of molecular phylogenetic and physiological approaches to study species from various locations could potentially resolve this question.

The physiological properties of *'Bangia'* species were recently investigated, specifically the response to heat stress. For instance, heat-inducible asexual reproduction to produce gametophytic clones via the release of asexual spores from gametophytes was observed in *'Bangia'* sp. ESS1 and other species collected from various locations [13–15]. In addition, *'Bangia'* sp. ESS1 can remember heat stress, allowing it to survive subsequent exposure to lethal temperatures after priming via exposure to sublethal temperatures [16]. This indicates that these algae maintain thermotolerance as a heat stress-responsive physiological state during subsequent non-stress control conditions, which increases the threshold level of heat-stress sensing. These findings suggest that *'Bangia'* sp. ESS1 responds to heat stress by producing new generations via accelerated asexual reproduction and the acquisition of thermotolerance based on stress memory. However, it is currently unknown whether these strategies are specific to *'Bangia'* 2, especially *'Bangia'* sp. ESS1, or whether they are common to all *'Bangia'* species.

Here, to address this question, we identified the marine species living at the rocky coast *'Bangia'* sp. ESS2, a Bangiales belonging to *'Bangia'* 3, and characterized its heat-stress responses in terms of growth-limiting temperature, accelerated asexual reproduction, and the acquisition of heat-stress tolerance compared to *'Bangia'* sp. ESS1 [12,16] and the freshwater species living close to the mountain stream *B. atropurpurea* [17]. Our results uncover the diverse heat-stress response strategies among *'Bangia'* species from different *'Bangia'* clades.

2. Results and Discussion

2.1. Identification of a Species Belonging to 'Bangia' 3

We amplified and sequenced a DNA fragment corresponding to the *rbcL* gene from gametophytes of a species collected on Kamomejima Island, Esashi, in April 2018 (GenBank accession number LC602264). We then performed phylogenetic analysis using the *rbcL* sequences from this species and species belonging to *'Bangia'* 2, including *'Bangia'* sp. ESS1 [12] and *'Bangia'* sp. OUCPT-01 [18], as well as the *'Bangia'* 3 clade of Bangiales [10]. As shown in Figure 1, the species was classified as a member of *'Bangia'* 3, with the closest relationship to *Bangia* sp. collected on Disko Island in Greenland, Rankin Inlet in Canada, and Chaichei Island in the United States, whose *rbcL* sequences were identical and deposited in GenBank under accession number AF043366 [19]. We designated the new species *'Bangia'* sp. ESS2 (ESS represents Esashi).

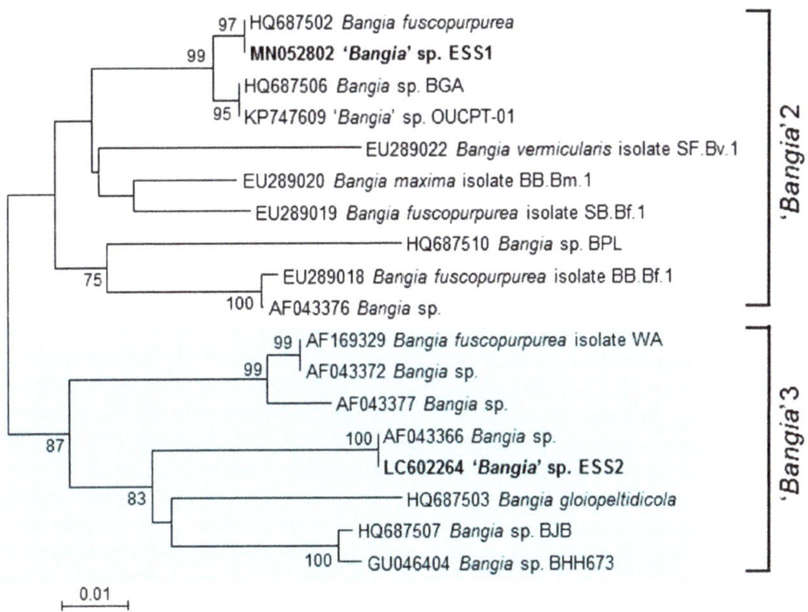

Figure 1. Phylogenetic identification of *'Bangia'* sp. ESS2. The phylogenetic tree was constructed by the neighbor-joining method using sequences of *rbcL* genes from different species from the *'Bangia'* 2 and *'Bangia'* 3 groups identified by Sutherland et al. [10]. The DDBJ/EMBL/GenBank accession numbers of the *rbcL* gene sequences are shown in front of the species names. *'Bangia'* sp. ESS1 and *'Bangia'* sp. ESS2 are highlighted. Bootstrap values over 50% from 1000 replicates are indicated at the nodes. Bar, 0.01 substitutions per site.

Bangia gloiopeltidicola, a well-known epiphytic seaweed of the red alga *Gloiopeltis furcata* (family Endocladiaceae), is a typical *'Bangia'* 3 species. However, *'Bangia'* sp. ESS2 adheres to rocks in the intertidal zone, pointing to the diversity of lifestyle strategies among *'Bangia'* 3 members.

2.2. Morphological and Developmental Properties of 'Bangia' sp. ESS2

Thalli of *'Bangia'* sp. ESS2 (Figure 2A) were usually uniseriate or biseriate filaments containing cylindrical vegetative cells (Figure 2B,G, respectively). The uniseriate filaments were 10.58 ± 0.85 µm in diameter ($N = 20$) and 8.65 ± 1.52 µm long ($N = 26$), and the biseriate filaments were 20.87 ± 2.12 µm in diameter ($N = 20$) and 6.445 ± 1.52 µm long ($N = 26$). Asexual spores, which are called monospores or archeospores, were released from both uniseriate and multiseriate thalli. When asexual spores were released from uniseriate thalli, the vegetative cells developed into asexual spores in the thalli and were released by rupture of the cell wall (Figure 2C,D). Release of asexual spores from multiseriate thalli required that uniseriate thalli develop into multiseriate thalli also known as asexual sporangia (Figure 2E–J). In this process, the vegetative cells grew to approximately twice as wide (Figure 2E) as uniseriate gametophytes (Figure 2B), and vertical cell division occurred, leading to the formation of a biseriate filament (Figure 2F,G). Subsequently, the number of vegetative cells in the filaments increased due to cell division, while the size of the cells decreased, resulting in the generation of asexual sporangia (Figure 2H). The asexual spores were released from the tip of the sporangium or developed into gametophytes without being released, to form gametophytic clones (Figure 2I,J).

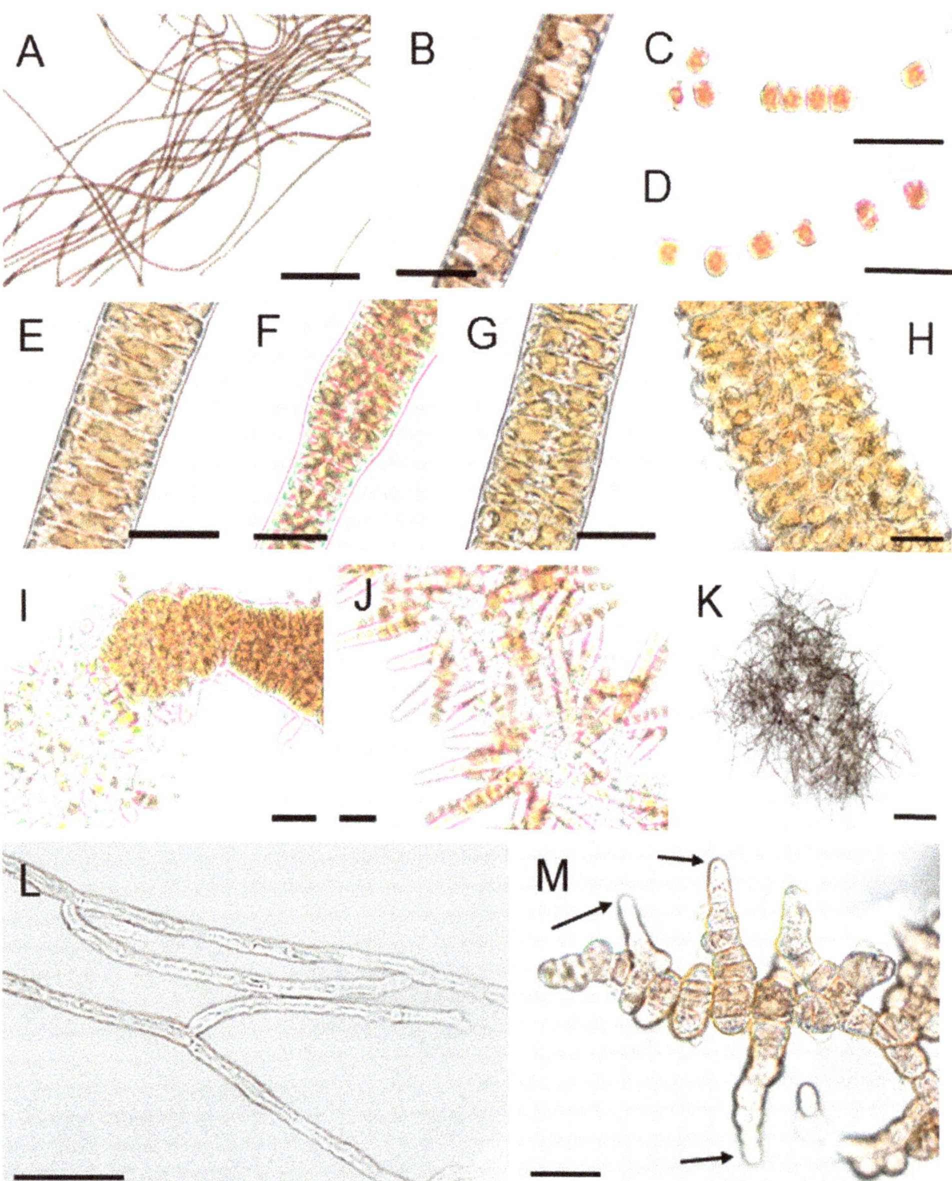

Figure 2. Morphological characteristics of thalli, conchocelis filaments, and conchosporangia of *'Bangia'* sp. ESS2: (**A**,**B**) Filamentous structure of gametophytic thalli; (**C**,**D**) Release of asexual spores from uniseriate thalli; (**E**–**H**) Asexual sporangium development, including doubling of the width of the vegetative cells, vertical cell division to form biseriate thalli, and random cell division to form multiseriate thalli; (**I**) Release of asexual spores from the tip of a multiseriate thallus; (**J**) Development of gametophytic thalli on a parental thallus without spore release; (**K**) Aggregate of conchocelis filaments, which appear black; (**L**) Magnified view of conchocelis filaments with branches; (**M**) Conchosporangia produced on conchocelis filaments. Many branches were produced in conchosporangia, unlike conchocelis filaments, which formed branches only infrequently; the pointed tip cells are indicated by arrows. Scale bars: 25 µm for A–J and M, 200 µm for K, 50 µm for L.

It was difficult to induce male and female gamete development under laboratory culture conditions. However, some naturally harvested thalli had already formed carposporangia via the fertilization of male and female gametes. Thus, these carposporangia produced conchocelis filaments in the laboratory (Figure 2K), which appeared black. Thus, it is clear that *'Bangia'* sp. ESS2 undergoes both sexual and asexual propagation during its life cycle. As shown in Figure 2L, all conchocelis filaments were uniseriate, with cylindrical cells (4.62 ± 0.72 µm in diameter (N = 20)/12.76 ± 1.65 µm long (N = 27)), from which branches were often produced. Conchosporangia developed on the conchocelis filaments as thick filaments composed of cells 16.97 ± 2.97 µm in diameter (N = 27) and 15.37 ± 2.69 µm long (N = 26), which underwent branching (Figure 2M). These findings indicate that the morphologies of *'Bangia'* sp. ESS2 are similar to those of other previously reported *'Bangia'* species [13,14].

Notably, the tip of each conchosporangium was pointed (Figure 2M), which is similar to the pointed conchosporangium tips of a *Porphyra* species collected in New Zealand [20]; however, other known *Neopyropia* species such as *N. pseudolinearis* and *N. yezoensis* have rounded tips [21,22]. Thus, as mentioned in Knight and Nelson [20], it appears that the morphology and shape of conchosporangia are effective taxonomic characteristics for discovering new species of Bangiales, although molecular validation via phylogenetic analysis would provide indispensable confirmatory evidence. Indeed, the conchosporangium tips of *'Bangia'* sp. collected in Fukaura, Aomori in Japan were pointed (see Figure 1C in [13]), suggesting that this species might be a *'Bangia'* 3 species, like *'Bangia'* sp. ESS2.

2.3. Growth-Limiting Temperatures of Gametophytic Thalli in 'Bangia' sp. ESS2 and Bangia Atropurpurea

When the thalli of *'Bangia'* sp. ESS2 were incubated at 15, 20, 25, and 28 °C, they appeared dark red-brown. These thalli could not be stained with erythrosine (Figure 3A), and most vegetative cells were alive (Figure 3B and Table S1). In addition, their survival was not affected by a 3-day incubation at 30 °C or a 1-day incubation at 32 °C (Figure 2B). However, a 7-day incubation at 30, 32, or 34 °C promoted greening and staining of the thalli with erythrosine (Figure 3A), indicating the death of thalli. Indeed, viability gradually decreased depending on the duration of incubation, and over 95% of cells were dead after 7 days of culture (Figure 3B and Table S1). These results are different from our previous findings for *'Bangia'* sp. ESS1 of the *'Bangia'* 2 clade [12,23], which cannot survive at 32 °C, whereas 80 and 40% survival were observed following incubation at 30 °C for 7 days and 3 weeks, respectively [15].

By contrast, *B. atropurpurea* thalli were sensitive to temperatures >20 °C: 70–80% of thalli survived a 7-day incubation at 20, 25, and 28 °C but not at 30, 32, or 34 °C (Figure 4 and Table S3). These results indicate that *B. atropurpurea* is more sensitive to heat stress than *'Bangia'* sp. ESS2 and that the level of intrinsic tolerance to heat stress in *'Bangia'* sp. ESS1 is highest among the three species, although the growth-limiting temperatures of *'Bangia'* sp. ESS1 and *'Bangia'* sp. ESS2 are similar and slightly higher than that of *B. atropurpurea*. Therefore, sensitivity to heat stress varied among species from different *'Bangia'* clades.

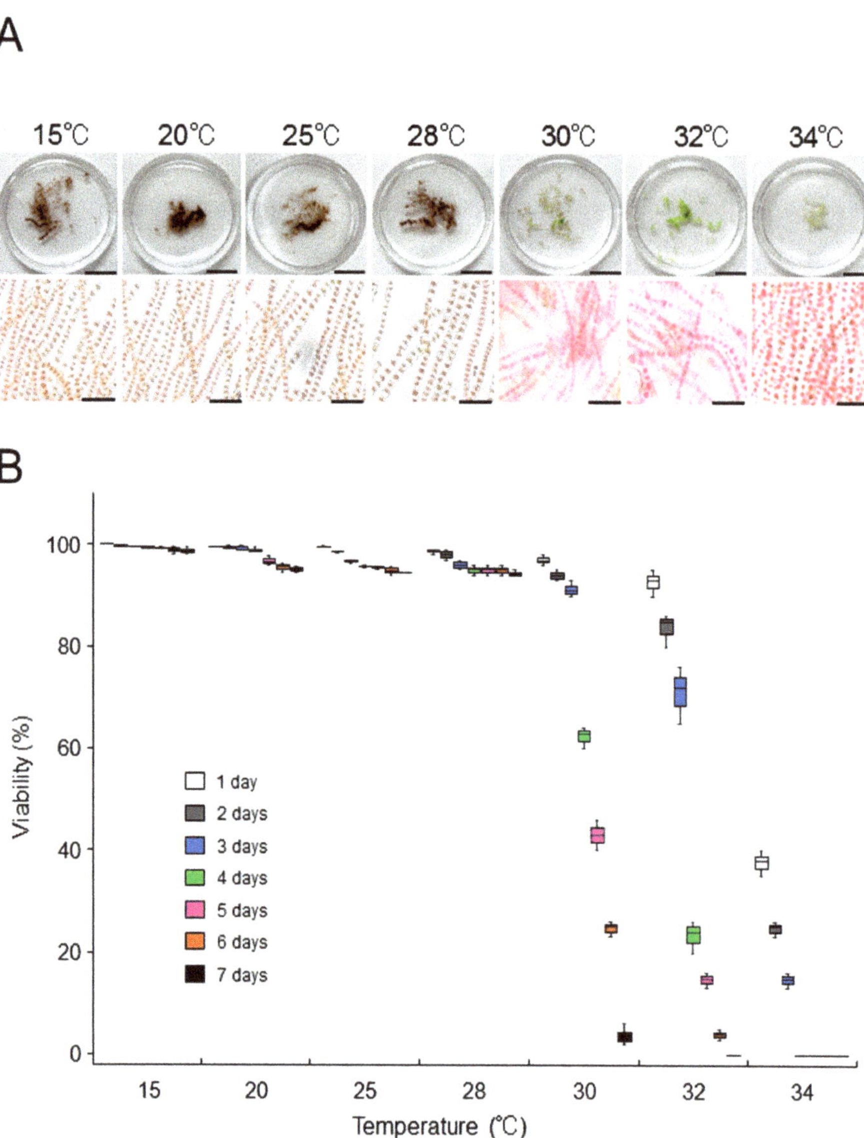

Figure 3. Effects of temperature on the viability of *'Bangia'* sp. ESS2 thalli. Samples of the laboratory-maintained culture (0.05 g) were incubated at 15, 20, 25, 28, 30, 32, and 34 °C for 7 days, and changes in body color and viability of cells were observed: (**A**) Comparison of body color (upper panels) and staining pattern with erythrosine (lower panels) among thalli treated with various temperatures for 7 days. Scale bars, 1 cm in the upper panels and 50 μm in the lower panels; (**B**) Quantification of viability. Viability of thalli incubated at various temperatures was examined daily by staining with 0.01% erythrosine. Error bars indicate the standard deviation of triplicate experiments (N = 3).

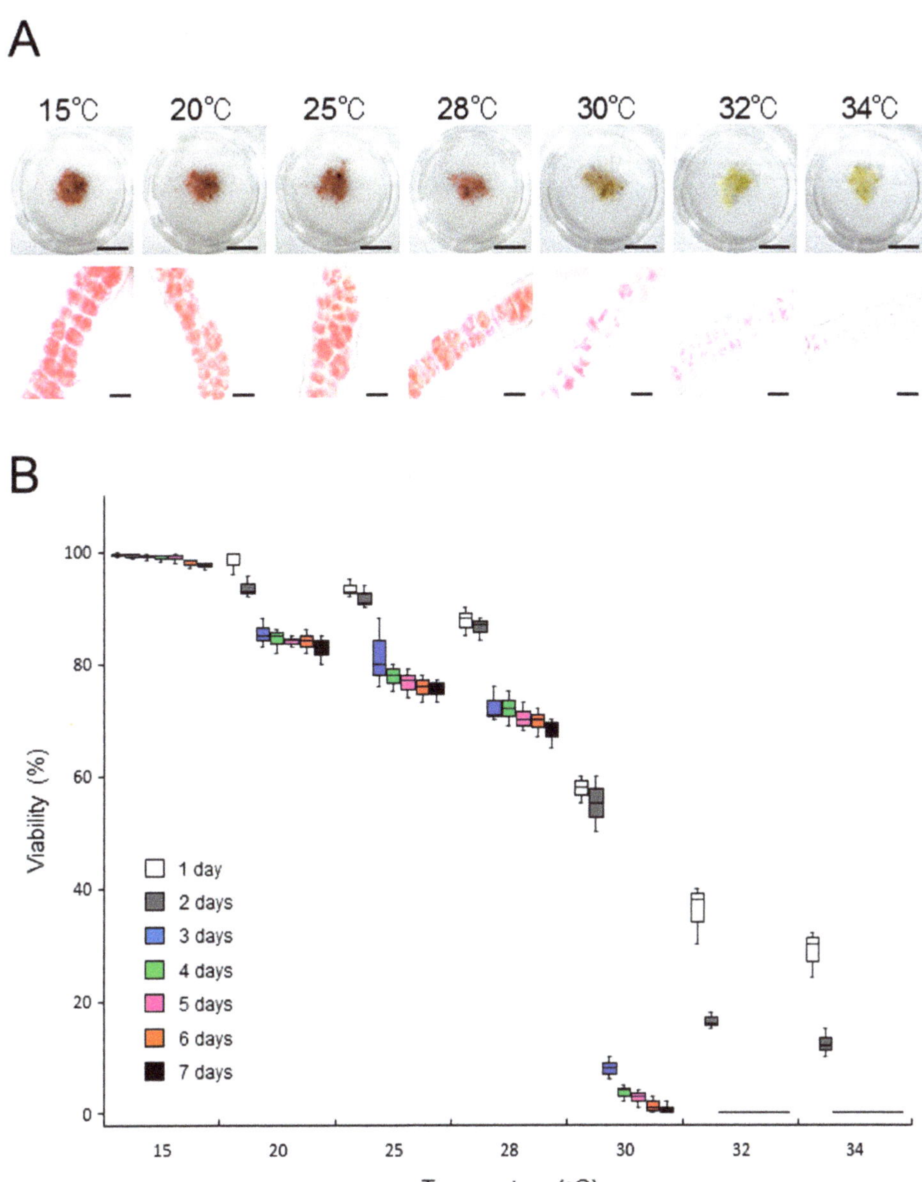

Figure 4. Effects of temperature on the viability of *Bangia atropurpurea* thalli. Samples of the laboratory-maintained culture (0.05 g) were incubated at 15, 20, 25, 28, 30, 32, and 34 °C for 7 days, and changes in body color and viability of cells were observed: (**A**) Comparison of body color (upper panels) and staining pattern with erythrosine (lower panels) among thalli treated with various temperature for 7 days. Scale bars, 1 cm in the upper panels and 25 µm in the lower panels; (**B**) Quantification of viability. Viability of thalli incubated at various temperatures was examined daily by staining with 0.01% erythrosine. Error bars indicate the standard deviation of triplicate experiments ($N = 3$).

2.4. Repression of the Asexual Life Cycle by Heat Stress

Since we previously observed the promotion of asexual sporulation at sublethal temperatures (such as 25 and 28 °C) in *'Bangia'* sp. ESS1 [15], we examined the effects of heat stress on asexual spore release in *'Bangia'* sp. ESS2 and *B. atropurpurea*. In *'Bangia'* sp. ESS2, the maximum release of asexual spores was observed 4 and 5 days after starting static culture of thalli at 15 °C, whereas increasing the culture duration reduced the number of released spores (Figure 5 and Table S3). Unexpectedly, heat stress repressed the release of asexual spores. Asexual sporulation by 5-day cultures gradually decreased with increasing temperature, and no release was observed at 32 or 34 °C (Figure 5 and Table S3). Thus, the release of asexual spores in *'Bangia'* sp. ESS2 is transient, with a peak at 4 and 5 days of static culture, and is repressed by heat treatment (Figure 5 and Table S3). By contrast, asexual reproduction was not observed in *B. atropurpurea* under heat-stress conditions.

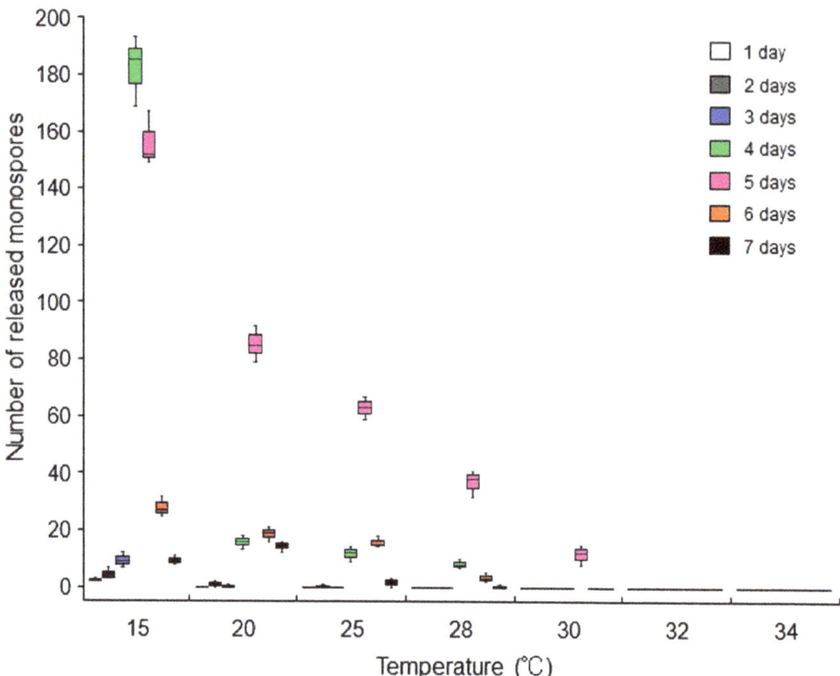

Figure 5. Effects of the heat stress on asexual propagation in *'Bangia'* sp. ESS2. Release of asexual spores from thalli (0.02 g) cultured statically at 15, 20, 25, 28, 30, 32, and 34 °C was counted every day for 7 days. Error bars indicate the standard deviation of triplicate experiments ($N = 3$).

The data in Figure 5 suggest that *'Bangia'* sp. ESS2 responds to the loss of water movement (calm stress), which means a loss of hydrodynamic stress via the static culture in dishes without water moving, by accelerating spore release via a heat-repressive pathway, which is different from heat stress-enhanced spore release in *'Bangia'* sp. ESS1 [15]. In fact, in *'Bangia'* sp. ESS1, calm stress itself had only a small effect on promoting asexual sporulation [15]; however, exposure to calm stress and subsequent freezing stress highly stimulated spore release after thawing [24]. Therefore, the calm stress-dependent release of asexual spores is basically conserved between *'Bangia'* sp. ESS1 and *'Bangia'* sp. ESS2, although the roles of calm stress in regulating asexual spore release in these two species differ. In addition, *'Bangia'* sp. ESS2 does not undergo heat stress-inducible sporulation (Figure 4 and Table S2), unlike *'Bangia'* sp. ESS1 [15]. Based on these findings, we propose that the strategies for stress-dependent resetting of the timing of the asexual lifecycle are different between the *'Bangia'* 2 and *'Bangia'* 3 clades of Bangiales. Why the strategies for asexual spore release differ between *'Bangia'* sp. ESS1 and *'Bangia'* sp. ESS2, and how heat and calm stress differentially promote the asexual life cycle in a species-dependent manner remain to be elucidated.

2.5. Defect in the Acquisition of Heat-Stress Tolerance

'Bangia' sp. ESS1 can acquire heat-stress tolerance by priming via incubation at non-lethal high temperatures, resulting in survival under subsequent exposure to lethal heat stress [16]. Thus, we addressed whether *'Bangia'* sp. ESS2 and *B. atropurpurea* are also able to acquire heat-stress tolerance. We pre-incubated *'Bangia'* sp. ESS2 thalli (Figure 6B,C) at 28 °C for 7 days, followed by incubation at 32 °C for 1 to 7 days (Figure 6A). Although the thalli were alive during pre-incubation at 28 °C (Figure 6D,E), as indicated in Figure 3, incubation at 32 °C for only 1 day killed all vegetative cells in the thalli (Figure 6F,G). These results demonstrate that *'Bangia'* sp. ESS2 cannot acquire thermotolerance under sublethal heat-stress conditions.

We then tested the ability of *B. atropurpurea* to acquire thermotolerance. Although this alga showed approximately 70% survival at 28 °C, but not 32 °C after 7 days of culture (Figures 4 and 7), the cultures were primed by treatment at 28 °C for 7 days, and the survival rate at the normally lethal temperature of 32 °C increased; for instance, a survival rate of 50% was observed after 7 days of culture at 32 °C (Figure 7, Table S4). Thus, *B. atropurpurea* can acquire thermotolerance by priming at 28 °C. We therefore examined whether *B. atropurpurea* can establish heat-stress memory to survive subsequent lethal high-temperature conditions. When the thalli were returned to 15 °C for 2 days after 7-day priming at 28 °C, incubation at 32 °C resulted in the loss of viability (Figure 7, Table S4). The results indicate that although *B. atropurpurea* has the ability to acquire thermotolerance by priming, it cannot establish heat-stress memory to maintain thermotolerance.

Since the living conditions of sessile organisms usually fluctuate dramatically and are often recurrent, the acquisition of thermotolerance via heat-stress memory (following exposure to sublethal heat-stress conditions) is thought to be essential for survival under subsequent lethal high-temperature stress in seaweeds including *'Bangia'* sp. ESS1, as well as terrestrial plants [16,25–28]. Thus, the inability to acquire heat-stress tolerance in *'Bangia'* sp. ESS2 and heat-stress memory in *B. atropurpurea* was unexpected; these are notable characteristics of poikilotherms. *'Bangia'* sp. ESS1 and *'Bangia'* sp. ESS2 were collected in May and April from different regions of the same island ([23] and see Section 3); thus, the environmental conditions experienced by these two species were likely similar. We propose that the different heat-stress response strategies of these species, rather than seasonal preferences, enable their compartmentalization on Kamomejima Island. In addition, *B. atropurpurea* was harvested from the rocky bed of a river (see Section 3); in such an environment, algae are usually splashed by river currents in the mountains and environmental conditions are relatively constant. Therefore, it is currently unclear whether the differences in the seasonal appearance of these species based on environmental conditions and their distinct heat-stress response strategies are related. Alternatively, since these three species belong to phylogenetically separated clades of *'Bangia'*, the variation in their heat-stress response strategies could be due to their different phylogenetic positions.

Despite these remaining unresolved questions, our results clearly indicate that species from different clades of *'Bangia'* employ different heat-stress response strategies. Specifically, *'Bangia'* sp. ESS2 and *B. atropurpurea* lack the intrinsic ability to acquire heat-stress tolerance and establish heat-stress memory to cope with recurrent changes in environmental conditions. Since this does not fit the general notion that poikilotherms require heat-stress tolerance and heat-stress memory [16,25–28], the presence of diverse responses to heat stress appears to be a special characteristic of *'Bangia'* species. Therefore, it remains to be elucidated why *'Bangia'* species from different clades employ different strategies in response to heat stress. Such information might help confirm the recently revised taxonomy of Bangiales [10].

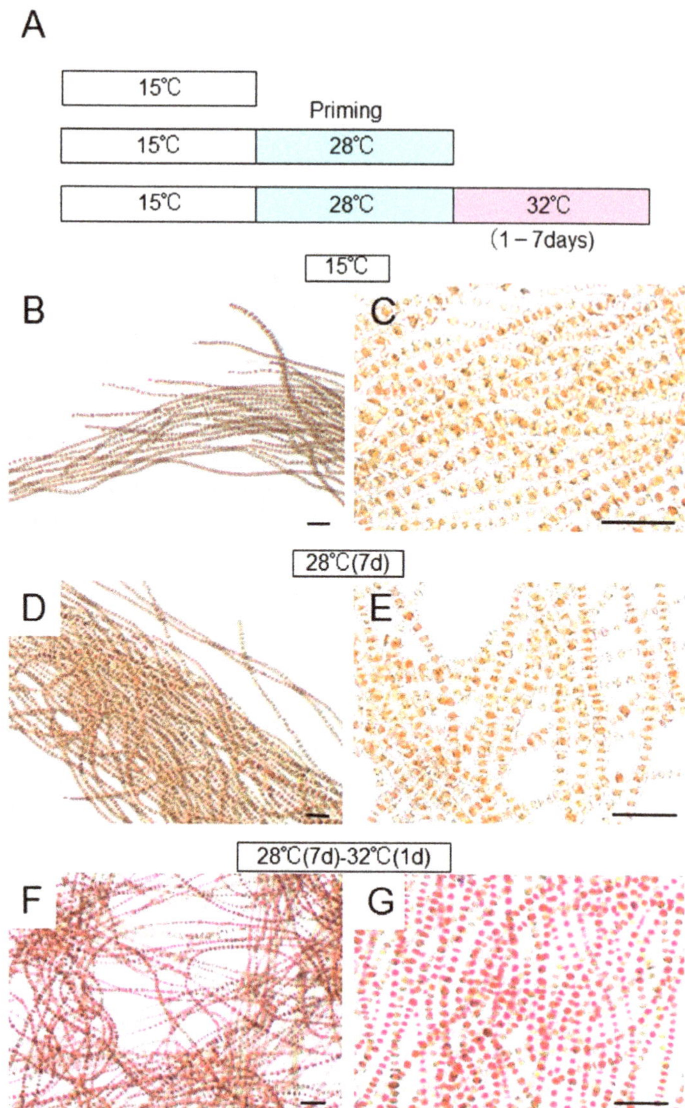

Figure 6. *'Bangia'* sp. ESS2 thalli fail to acquire heat-stress tolerance: (**A**) Schematic representation of the experimental design to assess the ability of *'Bangia'* sp. ESS2 to acquire heat-stress tolerance. Three temperature treatments were employed: control cultured at 15 °C, priming at 28 °C for 7 days, and treatment at 32 °C after priming for various durations from 1 to 7 days; (**B**,**C**) Photographs of control thalli grown at 15 °C shown at different magnifications; (**D**,**E**) Photographs of thalli primed at 28 °C for 7 days; (**F**,**G**) Photographs of thalli grown at 28 °C for 7 days, followed by 32 °C for 1 day. All thalli were stained with 0.01% erythrosine to visualize dead cells, which appear pink. Scale bars, 50 μm.

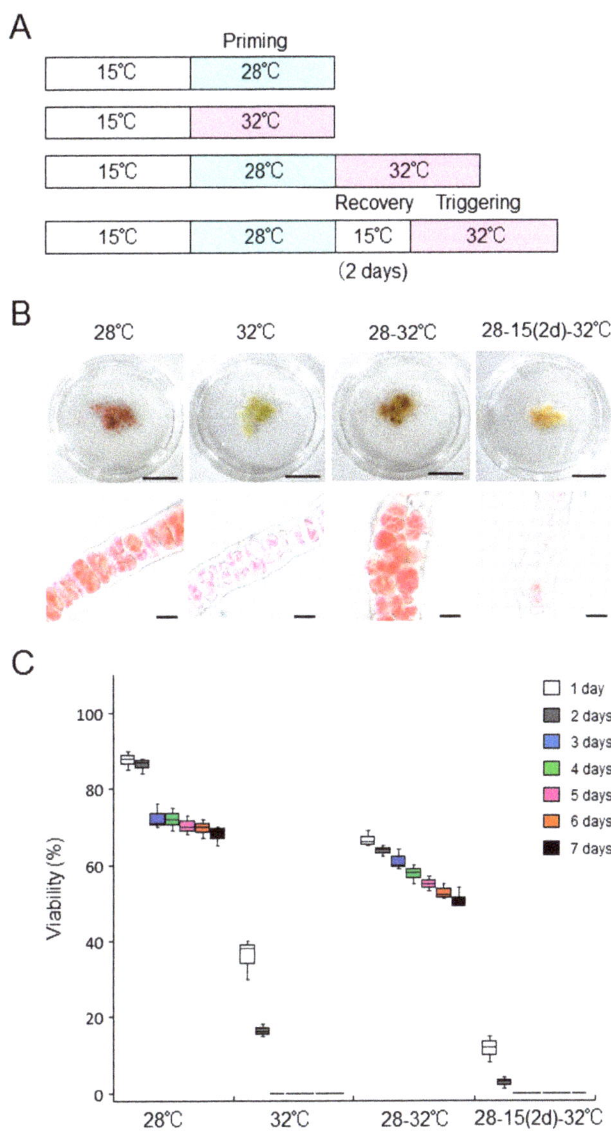

Figure 7. *Bangia atropurpurea* thalli fail to establish heat-stress memory: (**A**) Schematic representation of the experimental design to assess the ability of '*Bangia*' sp. ESS2 to memorize heat stress. Four temperature treatments were employed: priming at 28 °C for 7 days, direct transfer to 32 °C from 15 °C, incubation at 32 °C for 7 days after priming at 28 °C, and treatment at 32 °C after recovery for 2 days; (**B**) Comparison of body color (upper panel) and staining pattern with erythrosine (lower panel) among thalli treated with various temperature conditions, as indicated. All thalli were stained with 0.01% erythrosine to visualize dead cells, which appear pink. Scale bars, 1 cm and 25 μm for upper and lower panels, respectively; (**C**) Quantification of viability. Viability of thalli incubated at various temperature conditions was visualized daily by staining with 0.01% erythrosine. Error bars indicate the standard deviation of triplicate experiments ($N = 3$).

3. Materials and Methods

3.1. Algal Materials, Culture Conditions, and Morphological Observation

Gametophytes of filamentous *Bangia* grown on rocks were harvested on April 20, 2018 from Kamomejima Island (41°52′ N, 140°06′ E) in Esashi, Hokkaido in Japan. The thalli were maintained in sterilized artificial seawater (SEALIFE, Marinetech, Tokyo, Japan) enriched with ESS2 [29] under 60–70 µmol photons m^{-2} s^{-1} light with a short-day photoperiod (10 h light/14 h dark) at 15 °C with air filtered through a 0.22 µm filter (Whatman, Maidstone, UK). Conchocelis filaments appeared during the culture of thalli and were maintained as described for thalli; the conchosporangia parasitically developed on the conchocelis filaments. The culture medium was changed weekly. Thalli of the freshwater species *Bangia atropurpurea*, which were collected from the rocky bed of a river with a rapid current in Higashi-kawachisawa, Shizuoka, Japan in May 2005 and April 2006, were maintained according to Yokono et al. [17] in commercially available Ca^{2+}-rich mineral water (Contrex®, Nestlé Waters Marketing & Distribution) in plastic culture vessels, except that the culture conditions described above for marine *Bangia* species were utilized. Thalli, conchocelis, and conchosporangia were observed and imaged under an Olympus IX73 light microscope (Olympus, Tokyo, Japan) equipped with an Olympus DP22 camera.

3.2. Phylogenetic Analysis

Total genomic DNA was extracted from the species collected at Esashi from air-dried samples using a DNeasy Plant Mini kit (Qiagen, Valencia, CA, USA) according to the manufacturer's instructions. A 466-bp portion of the *rbcL* gene was amplified from the Esashi species with gene-specific primers (5′-AAGTGAACGTTACGAATCTGG-3′ and 5′-GATGCTTTATTTACACCCT-3′; [30]) using Ex Taq polymerase (TaKaRa Bio, Kusatsu, Japan) and sequenced on an ABI Model 3130 Genetic Analyzer (Life Technologies, Carlsbad, CA, USA). The nucleotide sequence of the amplified DNA fragment was deposited in DDBJ/EMBL/GenBank under accession number LC602264. A neighbor-joining phylogenetic tree was constructed with MEGA 7 software (https://www.megasoftware.net (accessed on 2 April 2021) using ClustalW to align the *rbcL* sequences from other *Bangia* species [10]. The accession numbers of these *rbcL* sequences are listed in front of the species' names in the phylogenetic tree.

3.3. Determining the Growth-Limiting Temperature

Each 0.05 g sample (fresh weight) of thalli (from aeration cultures grown at 15 °C) was incubated statically in dishes (Azunoru dish; 90 mm diameter × 20 mm height, As One Co., Ltd.) containing 50 mL of seawater for marine species or Contrex® for freshwater species at 20, 25, 28, 30, 32, and 34 °C for 7 days, while control experiments were performed at 15 °C. The viability of these thalli was visualized daily by staining with artificial seawater containing 0.01% erythrosine (Wako Pure Chemical Industries, Japan) as described by Kishimoto et al. [16]. In brief, thalli were stained for 5 min at room temperature, gently rinsed with artificial seawater or Contrex® to remove excess erythrosine, and mounted on a slide with medium. The thalli were observed and photographed under an Olympus IX73 light microscope equipped with an Olympus DP22 camera. Cells stained by the dye were defined as dead cells. Viability was calculated from the number of living and dead cells obtained using micrographs. Analysis of samples under each treatment condition was repeated three times.

3.4. Quantification of Released Asexual Spores

Each 0.02 g sample (fresh weight) of thalli (aeration-cultured at 15 °C) was incubated statically in dishes (Azunoru dish; 90 mm diameter × 20 mm height, As One) containing 50 mL of seawater at 15, 20, 25, 28, 30, 32, and 34 °C for 7 days. The number of asexual spores released onto the bottom of the dishes was counted daily by observation under an Olympus IX73 light microscope. Analysis of samples under each treatment condition was repeated three times.

3.5. Confirmation of Acquisition of Thermotolerance

Each 0.05 g (fresh weight) of thalli aeration-cultured at 15 °C was incubated statically in dishes (Azunoru dish; 90 mm diameter × 20 mm height, As One) containing 50 mL of seawater at 28 °C for 7 days or 28 °C for 7 days plus subsequent treatment at 32 °C for 1 to 7 days. The viability of these thalli was examined as described above. Analysis of samples under each treatment condition was repeated three times.

3.6. Statistical Analysis

Values are indicated by ±SD from triplicate experiments. Two-way ANOVA followed by a Tukey–Kramer test was used for multiple comparisons, and significant differences were determined using a cutoff value of $p < 0.05$.

4. Conclusions

We identified '*Bangia*' sp. ESS2 as a member of '*Bangia*' 3 and compared its heat-stress response strategies with those of *B. atropurpurea* and '*Bangia*' sp. ESS1. Our analysis revealed diversity in the heat-stress response strategies among these three species in terms of asexual sporulation, the acquisition of thermotolerance, and the memorization of heat stress. These findings suggest that the intrinsic abilities to respond to and tolerate heat stress vary among species from different clades of '*Bangia*'. Physiological and molecular biological studies of the mechanisms regulating heat-stress responses and the memorization of heat stress in species from inter- and intra-'*Bangia*' clades could help explain why these strategies are diverse in '*Bangia*'. Such information would increase our understanding of the biology of '*Bangia*' species from different clades and could confirm the recently revised taxonomy of Bangiales.

Supplementary Materials: The following are available online at https://www.mdpi.com/article/10.3390/plants10081733/s1, Table S1: Viability of vegetative cells in '*Bangia*' sp. ESS2 thalli in response to different durations of incubation under various temperature conditions, Table S2: Viability of vegetative cells in *Bangia atropurpurea* thalli in response to different durations of incubation under various temperature conditions, Table S3: Number of asexual spores released from '*Bangia*' sp. ESS2 thalli in response to different durations of incubation under various temperature conditions, Table S4: Viability of vegetative cells in *Bangia atropurpurea* thalli associated with the acquisition of heat-stress tolerance and the memorization of heat stress.

Author Contributions: Conceptualization, K.M.; methodology, S.S. and K.M.; validation, H.V.K. and K.M.; formal analysis, H.V.K.; investigation, H.V.K., P.K., and S.S.; data curation, H.V.K. and P.K.; writing—original draft preparation and writing—reviewing and editing, K.M.; visualization, H.V.K. and P.K.; resources H.U. and A.M.; supervision and project administration, K.M. All authors have read and agreed to the published version of the manuscript.

Funding: This research received no external funding.

Institutional Review Board Statement: Not applicable.

Informed Consent Statement: Not applicable.

Data Availability Statement: Data are contained within the article.

Acknowledgments: We are grateful to Hiromi Nakao from Betsukai Fishery Cooperative, Shyoichi Itakura from the Hiyama Fisheries Extension Office, Yoshitomo Hamaya from the Esashi Branch of the Fisheries Co-operative Association of Hiyama, and Hidehiko Sasaki from the Main Branch of the Fisheries Co-operative Association of Hiyama for permission and support in collecting seaweed on Kamomejima Island in Esasi. We also thank Ryunosuke Irie for technical assistance in phylogenetic analysis, Chengze Li for the maintenance of '*Bangia*' sp. ESS2, and Yuji Hiwatashi for technical assistance in microscopic observation. Ho Viet Khoa was supported by Ministry of Education, Culture, Sports, Science and Technology of Japan. Puja Kumari is an International Research Fellow supported by JSPS with Postdoctoral Fellowship for Research in Japan (Standard). We also thank Hiroyuki Mizuta for permission of Ho Viet Khoa to study in the Miyagi University.

Conflicts of Interest: The authors declare no conflict of interest.

References

1. Milstein, D.; de Oliveira, M.C. Molecular phylogeny of Bangiales (Rhodophyta) based on small subunit rDNA sequencing Emphasis on Brazilian *Porphyra* species. *Phycologia* **2005**, *44*, 212–221. [CrossRef]
2. Brodie, J.; Mortensen, A.M.M.; Ramirez, M.E.; Russell, S.; Rinkel, B. Making the links: Towards a global taxonomy for the red algal genus *Porphyra* (Bangiales, Rhodophyta). *J. Appl. Phycol.* **2008**, *20*, 939–949. [CrossRef]
3. Müller, K.M.; Cannone, J.J.; Sheath, R.G. A molecular phylogenetic analysis of the Bangiales (Rhodophyta) and description of a new genus and species, *Pseudobangia kaycoleia*. *Phycologia* **2005**, *44*, 146–155. [CrossRef]
4. Neefus, C.D.; Mathieson, A.C.; Klein, A.S.; Teasdale, B.; Gray, T.; Yarish, C. *Porphyra birdiae* sp. nov. (Bangiales, Rhodophyta): A new species from the northwest Atlantic. *Algae* **2002**, *17*, 203–216. [CrossRef]
5. Lindstrom, S.C.; Fredericq, S. rbcL gene sequences reveal relationships among north-east Pacific species of *Porphyra* (Bangiales, Rhodophyta) and a new species, *P. aestivalis*. *Phycol. Res.* **2003**, *51*, 211–224. [CrossRef]
6. Brodie, J.; Bartsch, I.; Neefus, C.; Orfanidis, S.; Bray, T.; Mathieson, A.C. New insights into the cryptic diversity of the North Atlantic-Mediterranean '*Porphyra leucosticta*' complex: *P. olivii* sp. nov. and *P. rosengurttii* (Bangiales, Rhodophyta). *Eur. J. Phycol.* **2007**, *42*, 3–28. [CrossRef]
7. Lindstrom, S.C. Cryptic diversity, biogeography and genetic variation in northeast Pacific species of *Porphyra* sensu lato (Bangiales, Rhodophyta). *J. Appl. Phycol.* **2008**, *20*, 951–962. [CrossRef]
8. Niwa, K.; Iida, S.; Kato, A.; Kawai, H.; Kikuchi, N.; Kobiyama, A.; Aruga, Y. Genetic diversity and introgression in two cultivated species (*Porphyra yezoensis and Porphyra tenera*) and closely related wild species of *Porphyra* (Bangiales, Rhodophyta). *J. Phycol.* **2009**, *45*, 493–502. [CrossRef] [PubMed]
9. Niwa, K.; Kikuchi, N.; Hwang, M.-S.; Choi, H.-G.; Aruga, Y. Cryptic species in the *Pyropia yezoensis* complex (Bangiales, Rhodophyta): Sympatric occurrence of two cryptic species even on same rocks. *Phycol. Res.* **2014**, *62*, 36–43. [CrossRef]
10. Sutherland, J.; Lindstrom, S.; Nelson, W.; Brodie, J.; Lynch, M.; Hwang, M.S.; Choi, H.G.; Miyata, M.; Kikuchi, N.; Oliveira, M.; et al. A new look at an ancient order: Generic revision of the Bangiales. *J. Phycol.* **2011**, *47*, 1131–1151. [CrossRef] [PubMed]
11. Sánchez, N.; Vergés, A.; Peteiro, C.; Sutherland, J.E.; Brodie, J. Diversity of bladed Bangiales (Rhodophyta) in western Mediterranean: Recognition of the genus Themis and descriptions of *T. ballesterosii* sp. nov., *T. iberica* sp. nov., and *Pyropia parva* sp. nov. *J. Phycol.* **2014**, *50*, 908–929, Corrigendum in **2015**, *46*, 206. [CrossRef]
12. Li, C.; Irie, R.; Shimada, S.; Mikami, K. Requirement of different normalization genes for quantitative gene expression analysis under abiotic stress conditions in '*Bangia*' sp. ESS1. *J. Aquat. Res. Mar. Sci.* **2019**, *2019*, 194–205.
13. Notoya, M.; Iijima, N. Life history and sexuality of archeospore and apogamy of *Bangia* atropurpurea (Roth) Lyngbye (Bangiales, Rhodophyta) from Fukaura and Enoshima, Jpn. *Fish. Sci.* **2003**, *69*, 799–805. [CrossRef]
14. Wang, W.-J.; Zhu, J.-Y..; Xu, P.; Xu, J.R.; Lin, X.Z.; Huang, C.K.; Song, W.L.; Peng, G.; Wang, G.C. Characterization of the life history of *Bangia fuscopurpurea* (Bangiaceae, Rhodophyta) in connection with its cultivation in China. *Aquaculture* **2008**, *278*, 101–109. [CrossRef]
15. Mikami, K.; Kishimoto, I. Temperature promoting the asexual life cycle program in *Bangia fuscopurpurea* (Bangiales, Rhodophyta) from Esashi in the Hokkaido Island, Japan. *Algal Resour.* **2018**, *11*, 25–32.
16. Kishimoto, I.; Ariga, I.; Itabashi, Y.; Mikami, K. Heat-stress memory is responsible for acquired thermotolerance in *Bangia fuscopurpurea*. *J. Phycol.* **2019**, *55*, 971–975. [CrossRef] [PubMed]
17. Yokono, M.; Uchida, H.; Suzawa, Y.; Akiomoto, S.; Murakami, A. Stabilization and modulation of the phycobilisome by calcium in the calciphilic freshwater red alga *Bangia atropurpurea*. *Biochim. Biophys. Acta* **2012**, *1817*, 306–3011. [CrossRef] [PubMed]
18. Cao, M.; Bi, G.; Mao, Y.; Li, G.; Kong, F. The first plastid genome of a filamentous taxon '*Bangia*' sp. OUCPT-01 in the Bangiales. *Sci. Rep.* **2018**, *8*, 10688. [CrossRef]
19. Müller, K.M.; Sheath, R.G.; Vis, M.L.; Crease, T.J.; Cole, K.M. Biogeography and systematics of *Bangia* (Bangiales, Rhodophyta) based on the Rubisco spacer, rbcL gene and 18S rRNA gene sequences and morphometric analyses. 1. North America. *Phycologia* **1998**, *37*, 195–207. [CrossRef]
20. Knight, G.A.; Nelson, W.A. An evaluation of characters obtained from life history studies for distinguishing New Zealand *Porphyra* species. *J. Appl. Phycol.* **1999**, *11*, 411–419. [CrossRef]
21. Kim, N.-G. Culture studies of *Porphyra dentate* and *P. pseudolinearis* (Bangia, Rhodophyta), two dioecious species from Korea. *Hydrobiologia* **1999**, *398/399*, 127–135. [CrossRef]
22. Takahashi, M.; Mikami, K. Phototropism in the marine red macroalga *Pyropia yezoensis*. *Am. J. Plant Sci.* **2016**, *7*, 2412–2428. [CrossRef]
23. Hirata, R.; Takahashi, M.; Saga, N.; Mikami, K. Transient gene expression system established in *Porphyra yezoensis* is widely applicable in Bangiophycean algae. *Mar. Biotechnol.* **2011**, *13*, 1038–1047. [CrossRef]
24. Omuro, Y.; Khoa, H.V.; Mikami, K. The absence of hydrodynamic stress promotes acquisition of freezing tolerance and freeze-dependent asexual reproduction in the red alga '*Bangia*' sp. ESS1. *Plants* **2021**, *10*, 465. [CrossRef]
25. Bruce, T.J.A.; Matthes, M.C.; Napier, J.A.; Pickett, J.A. Stressful "memories" of plants: Evidence and possible mechanisms. *Plant Sci.* **2007**, *173*, 603–608. [CrossRef]
26. Bäurle, I. Plant heat adaptation: Priming in response to heat stress. *F1000Res.* **2016**, *5*, 694. [CrossRef]
27. Crisp, P.A.; Ganguly, D.; Eichten, S.R.; Borevitz, J.O.; Pogson, B.J. Reconsidering plant memory: Intersections between stress recovery, RNA turnover, and epigenetics. *Sci. Adv.* **2016**, *2*, e1501340. [CrossRef]

28. Sanyal, R.P.; Misra, H.S.; Saini, A. Heat-stress priming and alternative splicing-linked memory. *J. Exp. Bot.* **2018**, *69*, 2431–2434. [CrossRef]
29. Li, C.; Ariga, I.; Mikami, K. Difference in nitrogen starvation-inducible expression patterns among phylogenetically diverse ammonium transporter genes in the red seaweed *Pyropia yezoensis*. *Am. J. Plant Sci.* **2019**, *10*, 1325–1349. [CrossRef]
30. Ohnishi, M.; Kikuchi, N.; Iwasaki, T.; Kawaguchi, R.; Shimada, S. Population genomic structures of endangered species (CR+EN), *Pyropia tenera* (Bangiales, Rhodophyta). *Jpn. J. Phycol. (Sorui)* **2013**, *61*, 87–96.

 plants 

Article

Vegetative Reproduction Is More Advantageous Than Sexual Reproduction in a Canopy-Forming Clonal Macroalga under Ocean Warming Accompanied by Oligotrophication and Intensive Herbivory

Hikaru Endo [1,2,*], Toru Sugie [1], Yukiko Yonemori [1], Yuki Nishikido [1], Hikari Moriyama [1], Ryusei Ito [3] and Suguru Okunishi [1]

1 Faculty of Fisheries, Kagoshima University, Kagoshima 890-0056, Japan; toru90118@gmail.com (T.S.); k5454895@kadai.jp (Y.Y.); k5945870@kadai.jp (Y.N.); k0531012@kadai.jp (H.M.); okunishi@fish.kagoshima-u.ac.jp (S.O.)

2 United Graduate School of Agricultural Sciences, Kagoshima University, Kagoshima 890-0065, Japan

3 Fisheries Research Division, Oita Prefectural Agriculture, Forestry and Fisheries Research Center, Bungotakada 879-0608, Japan; ito-ryusei@pref.oita.lg.jp

* Correspondence: h-endo@fish.kagoshima-u.ac.jp; Tel.: +81-99-286-4131

Abstract: Ocean warming and the associated changes in fish herbivory have caused polarward distributional shifts in the majority of canopy-forming macroalgae that are dominant in temperate Japan, but have little effect on the alga *Sargassum fusiforme*. The regeneration ability of new shoots from holdfasts in this species may be advantageous in highly grazed environments. However, little is known about the factors regulating this in *Sargassum* species. Moreover, holdfast tolerance to high-temperature and nutrient-poor conditions during summer has rarely been evaluated. In the present study, *S. fusiforme* holdfast responses to the combined effects of temperature and nutrient availability were compared to those of sexually reproduced propagules. The combined effects of holdfast fragmentation and irradiance on regeneration were also evaluated. Propagule growth rate values changed from positive to negative under the combination of elevated temperature (20 °C–30 °C) and reduced nutrient availability, whereas holdfasts exhibited a positive growth rate even at 32 °C in nutrient-poor conditions. The regeneration rate increased with holdfast fragmentation (1 mm segments), but was unaffected by decreased irradiance. These results suggest that *S. fusiforme* holdfasts have a higher tolerance to high-temperature and nutrient-poor conditions during summer than propagules, and regenerate new shoots even if 1-mm segments remain in shaded refuges for fish herbivory avoidance.

Keywords: climate change; foundation species; fucoid brown algae; non-additive effect; simulated herbivory

 check for updates

Citation: Endo, H.; Sugie, T.; Yonemori, Y.; Nishikido, Y.; Moriyama, H.; Ito, R.; Okunishi, S. Vegetative Reproduction Is More Advantageous Than Sexual Reproduction in a Canopy-Forming Clonal Macroalga under Ocean Warming Accompanied by Oligotrophication and Intensive Herbivory. *Plants* **2021**, *10*, 1522. https://doi.org/10.3390/plants10081522

Academic Editor: Koji Mikami

Received: 30 June 2021
Accepted: 21 July 2021
Published: 26 July 2021

Publisher's Note: MDPI stays neutral with regard to jurisdictional claims in published maps and institutional affiliations.

Copyright: © 2021 by the authors. Licensee MDPI, Basel, Switzerland. This article is an open access article distributed under the terms and conditions of the Creative Commons Attribution (CC BY) license (https://creativecommons.org/licenses/by/4.0/).

1. Introduction

Plant reproduction can be divided into sexual and asexual reproduction; the latter includes somatic embryogenesis and vegetative reproduction, which occurs without embryo formation [1]. Vegetative reproduction is common in clonal plants, which produce new shoots (i.e., ramets) from roots, stolons, or rhizomes [2–4]. The newly produced ramets obtain resources, such as nutrients and carbohydrates, from physiologically integrated mother plants through connections [4]. Moreover, fragmentation of these ramets via the disturbance or senescence of these connections often enhances new ramet production [2,3]. Therefore, vegetative reproduction may be more advantageous than sexual reproduction under resource-limited and highly disturbed environments.

Brown, red, and green macroalgae are the major primary producers in coastal marine ecosystems. Specifically, the canopy-forming large brown algae, kelp (Laminariales) and fucoid species (Fucales), are highly productive and act as foundation species [5], providing food, habitats, and spawning grounds for various marine organisms [6,7]. Furthermore, the conservation and restoration of marine macroalgal forests, which export carbon to the deep sea, may contribute to the mitigation of climate change caused by an increase in the atmospheric CO_2 level [8,9]. However, these algal forests have been declining due to ocean warming [10]. Above-average temperatures, combined with nutrient-poor conditions during the summer, have been known to cause physiological stress in macroalgal species [11–14].

Moreover, ocean warming has caused a range expansion of tropical herbivorous fishes into temperate waters, resulting in an increase in their grazing activity, especially in ocean warming hotspots, such as the Mediterranean and southern Japan [15]. Additionally, in the western North Pacific around southern Japan, nutrient concentrations in the surface mixed layer have been declining because the mixing of nutrient-poor surface water and nutrient-rich deep water has been suppressed by ocean warming or longer-term natural climate change [16]. Consequently, the majority of the kelp and fucoid species that are dominant in temperate Japan have shifted their distributional range towards the pole [17]. However, such poleward range shifts have not been observed in the fucoid *Sargassum fusiforme* [17], implying that this species might have reproductive traits that allow for its survival in warm, nutrient-poor, and highly grazed environments in southern Japan.

Sargassum species generally have perennial holdfasts (analogous to rhizoids), stipes (analogous to stems), and annual shoots (i.e., main branches), which show large seasonal variations in biomass and length, with the exception of annual species such as *S. horneri* [18–21]. In temperate *Sargassum* species, including *S. fusiforme*, these shoots commonly germinate from stipes during summer, grow between autumn and spring, and decay during the subsequent summer after the production of propagules via sexual reproduction [18,20]. Moreover, vegetative reproduction via the regeneration of new shoots from holdfasts has been reported in several *Sargassum* species, including *S. fusiforme* [22–24]. In southern Japan, where fish herbivory is intensive between summer and autumn, and is weaker during winter [25], *Sargassum* shoots derived from holdfasts or propagules only grow from winter to spring, and decay during summer [26–28]. Therefore, these propagules and holdfasts, rather than the shoots, are exposed to warm and nutrient-poor conditions during summer. Previous studies have shown the effect of increased temperature on propagule growth [29,30], and the combined effects of temperature and nutrient availability or salinity on shoot growth [31–33] in *Sargassum* species. However, the combined effects of elevated temperature and reduced nutrient availability on the growth of propagules and holdfasts have rarely been evaluated; therefore, it is unclear whether sexual or vegetative reproduction is more advantageous under warm and nutrient-poor environments.

Moreover, Ito et al. [22] showed that the regeneration of new shoots from holdfasts was enhanced by cutting the filamentous holdfasts into segments <2.5 mm in length in *S. fusiforme*. This implies that this species might regenerate even after the holdfasts are fragmented by fish herbivory, as reported in *S. swartzii* [24]. They also reported that the percentage of *S. fusiforme* holdfasts that regenerated new shoots tended to increase in response to elevations in temperature (from 17 °C to 23 °C) and irradiance (from 50 to 230 μmol photons m^{-2} s^{-1}) [22]. However, the effects of a broader range of temperatures and nutrient availability on regeneration have not been studied and therefore the most important factor regulating regeneration is unclear.

Furthermore, microtopographic refuges, such as crevices, are known to enhance the recruitment and survival of *Sargassum* propagules in tropical regions, where intensive fish herbivory occurs [34]. Although these microhabitats are predicted to act as holdfast refuges from fish herbivory, reduced light availability in a shaded crevice may antagonize the positive effect of holdfast fragmentation by fish herbivory. However, the combined effects of fragmentation and decreased irradiance on regeneration are unclear based on the results of single-factor studies.

Sargassum fusiforme is common between lower intertidal and upper subtidal reefs in Japan, China, and Korea [35,36]. This species is edible and has been cultivated in these countries. Due to its commercial importance, several studies on the ecological and physiological traits of this species have been conducted [20,22,33,37], although the effects of ocean warming, nutrients, and herbivory on its reproductive traits have not been examined. Novel knowledge of the reproductive traits of this species may improve seeding methods for its cultivation.

In the present study, four laboratory culture experiments of *S. fusiforme* were conducted to examine (1) the combined effects of temperature (10 °C–30 °C) and nutrient availability on propagule growth, (2) the combined effects of temperature (15 °C–30 °C) and nutrient availability on the growth and regeneration rates of holdfasts, (3) the effect of high temperature (30 °C–38 °C) on the growth and regeneration rates of holdfasts, and (4) the combined effects of holdfast fragmentation and irradiance on the growth and regeneration rates of holdfasts in this species.

2. Results

2.1. Experiment 1: Combined Effects of Temperature and Nutrients on Propagules

Mean (± standard deviation) dissolved inorganic nitrogen (DIN) concentrations in 10% Provasoli's enriched seawater (PESI), 5% PESI, and sterile natural seawater (SSW) were 106.36 ± 1.07 μM, 74.89 ± 7.63 μM, and 6.94 ± 0.17 μM, respectively. The two-way analysis of variance (ANOVA) detected significant effects of temperature and nutrients, and their interaction on the relative growth rate of propagules (Table 1). The results of Tukey's test showed that there was no significant difference in the growth rate among temperatures in 10% PESI treatments, whereas the values decreased in response to elevated temperature from 20 °C to 30 °C in the 5% PESI treatment (Figure 1). The growth rate was significantly lower in non-enriched SSW treatments than in 10% PESI treatments at all temperatures. Moreover, the positive growth rate became negative in response to temperature elevation from 20 °C to 30 °C in the SSW treatment.

Table 1. Results of two-way ANOVA on the effects of temperature and nutrient availability on the relative growth rate of *Sargassum fusiforme* germlings.

Source	df	MS	*F*	*p*	
Temperature (T)	2	1.385	32.408	<0.001	*
Nutrient (N)	2	2.420	56.647	<0.001	*
T × N	4	0.351	8.219	<0.001	*

* Statistical significance.

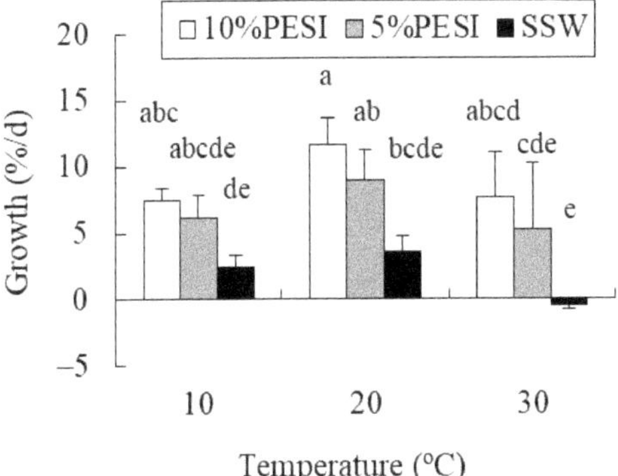

Figure 1. Relative growth rate of *Sargassum fusiforme* propagules cultured in nine different treatments (Mean + SD, $n = 6$). Different small letters indicate statistical significance among different treatments ($p < 0.05$).

2.2. Experiment 2: Combined Effects of Temperature and Nutrients on Holdfasts

The mean (± standard deviation) DIN concentrations in 25% PESI and SSW were 150.47 ± 1.18 μM and 5.26 ± 2.70 μM, respectively. The holdfast relative growth rate was significantly affected by temperature and nutrient availability, although their interaction was not significant (Table 2). The growth rate was significantly lower at 30 °C than at 20 °C, and lower in SSW treatments than in 25% PESI treatments (Figure 2). In contrast, the relative regeneration rate was unaffected by both temperature and nutrients, although values tended to decrease in response to elevated temperature from 20 °C to 30 °C, especially in the 25% PESI treatment.

Table 2. Results of two-way ANOVA on the effects of temperature and nutrient availability on the relative growth and regeneration rates of *Sargassum fusiforme* holdfasts.

Source	df	MS	*F*	*p*	
Growth rate					
Temperature (T)	3	0.915	3.887	0.016	*
Nutrient (N)	1	1.129	4.797	0.034	*
T × N	3	0.052	0.222	0.881	
Regeneration rate					
Temperature (T)	3	0.254	2.478	0.075	
Nutrient (N)	1	0.100	0.973	0.330	
T × N	3	0.045	0.441	0.725	

* Statistical significance.

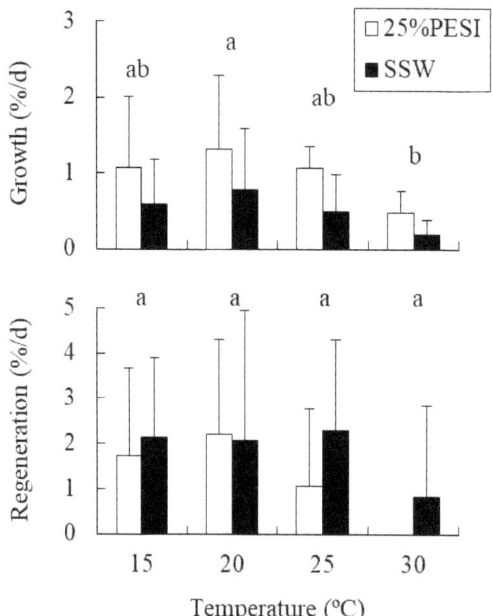

Figure 2. Relative growth and regeneration rates of *Sargassum fusiforme* holdfasts cultured in eight different treatments (Mean + SD, $n = 6$). Different small letters indicate statistical significance among different temperature treatments ($p < 0.05$).

2.3. Experiment 3: Effect of High Temperature on Holdfasts

A significant effect of temperature (between 30 °C and 38 °C) on the relative growth rate of holdfasts was detected by one-way ANOVA (df = 4, MS = 0.859, $F = 2.998$, $p = 0.040$). Tukey's test indicated that the growth rate was higher at 30 °C and 32 °C than at 36 °C under a significance level of $p < 0.1$ (Figure 3), although the differences among temperatures were not detected under a $p < 0.05$ level. The mean value was positive at 30 °C–32 °C and was negative at 34 °C–38 °C. In contrast, the relative regeneration rate was not significantly affected by temperature (df = 4, MS = 0.306, $F = 0.75$, $p = 0.567$). Regeneration was even observed at 30 °C and 34 °C, but not in any other treatments.

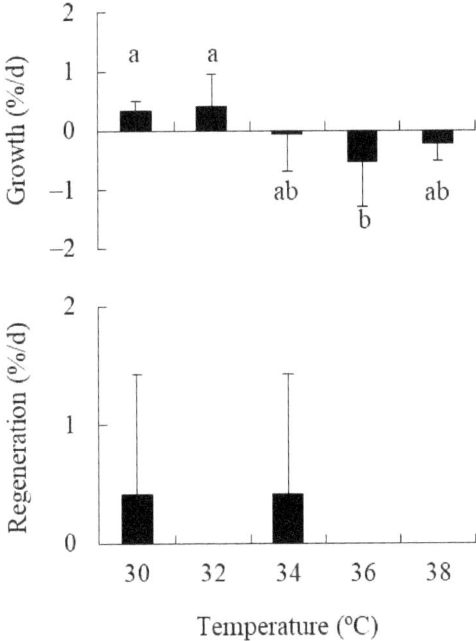

Figure 3. Relative growth and regeneration rates of *Sargassum fusiforme* holdfasts cultured in five different temperatures (Mean + SD, $n = 6$) using SSW as the culture media. Different small letters indicate statistical significance among different temperature treatments ($p < 0.1$).

2.4. Experiment 4: Combined Effects of Fragmentation and Irradiance on Holdfasts

The holdfast relative growth rate was significantly affected by irradiance, but not by fragmentation or their interaction (Table 3). Values decreased in response to reduced irradiance from 130 to 30 μmol photons m^{-2} s^{-1} (Figure 4). In contrast, the relative regeneration rate was significantly affected by fragmentation, but not by irradiance or their interaction (Table 3). Fragmentation significantly increased the regeneration rate, even in the low-irradiance treatments (Figure 4).

Table 3. Results of two-way ANOVA on the effects of fragmentation and irradiance on the relative growth and regeneration rates of *Sargassum fusiforme* holdfasts.

Source	df	MS	*F*	*p*	
Growth rate					
Fragmentation (F)	1	0.013	0.029	0.866	
Irradiance (I)	1	2.260	5.228	0.033	*
F × I	1	0.038	0.089	0.769	
Regeneration rate					
Fragmentation (F)	1	33.982	5.156	0.034	*
Irradiance (I)	1	0.187	0.028	0.868	
F × I	1	0.995	0.151	0.702	

* Statistical significance.

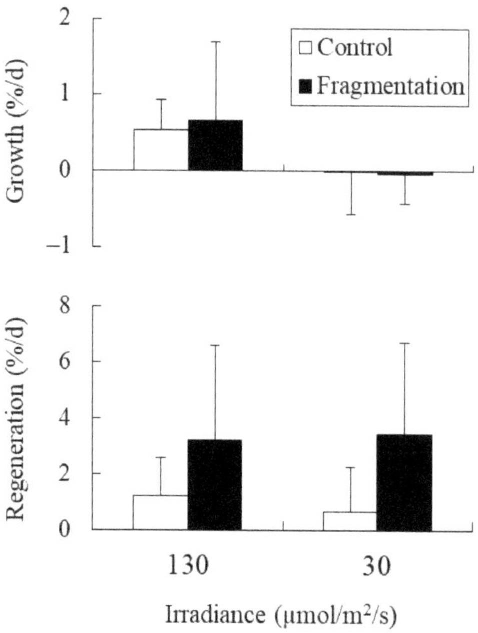

Figure 4. Relative growth and regeneration rates of *Sargassum fusiforme* holdfasts cultured in eight different treatments (Mean + SD, $n = 6$) at 30 °C using SSW as the culture media.

3. Discussion

Baba [29] examined the combined effects of seven temperature levels (10 °C, 15 °C, 20 °C, 25 °C, 30 °C, 32 °C, and 34 °C) and four irradiance levels (10, 25, 100, and 180 μmol photons m^{-2} s^{-1}) on the relative growth rate of *S. fusiforme* propagules in a 20-d experiment using 100% PESI (DIN = ca. 800 μM) as a culture medium. Growth rates were reported to be the highest at 25 °C–30 °C at 100–180 μmol photons m^{-2} s^{-1}. The present 21-d study evaluated the combined effects of three temperature levels (10 °C, 20 °C, and 30 °C) and three nutrient levels (10% PESI, 5% PESI, and SSW) on the growth rate of *S. fusiforme* propagules, and found a significant interaction between temperature and nutrient availability. Negative effects of increased temperature from 20 °C to 30 °C were not detected in 10% PESI treatments (DIN = ca. 100 μM), but were found in 5% PESI treatments (DIN = ca. 75 μM). Moreover, the growth rate in non-enriched SSW treatments (DIN = ca. 7 μM) was lower than those in 10% PESI treatments at all temperatures, and changed from positive to negative values in response to a temperature elevation from 20 °C to 30 °C. These results suggest that the high-temperature tolerance of *S. fusiforme* propagules strongly depends on nutrient availability. Similar results have been obtained in our previous studies using juvenile sporophytes of the kelps *Undaria pinnatifida*, *Ecklonia cava*, and *Sacchrina japonica* [12–14]. Therefore, the early life stages of kelp and fucoid species appear to be vulnerable to reduced nutrient availability, especially under warm conditions, probably because of their small size and limited resource accumulation.

Yatsuya et al. [23] reported that holdfasts of *Sargassum piluriferum* and *S. alternato-pinnatum* regenerated new shoots after incubation at 32.5 °C for 5–17 d. In the present study, there were significant effects of temperature and nutrient availability on the growth of *S. fusiforme* holdfasts, although there was no significant interaction between the two factors, indicating an additive effect. Holdfast growth decreased in response to elevated temperature from 20 °C to 30 °C and reduced nutrient availability from 25% PESI (DIN = ca.150 μM) to SSW (DIN = ca.5 μM). However, the growth rate remained at positive values even at 30 °C in nutrient-poor SSW treatment, in contrast to the results of the propagules. Moreover, the growth rates maintained positive values at 32 °C, and became negative at 34 °C in our subsequent 28-d experiment using SSW as a culture medium, whereas the propagules decreased their growth rate in response to the temperature elevation from 30 °C to 32 °C and withered within 4 d at 34 °C under nutrient-rich 100% PESI conditions [29]. These results suggest that the holdfasts of this species have a higher tolerance to high-temperature and nutrient-poor conditions during summer than propagules of the same species. High-temperature tolerance of marine macroalgae is known to be associated with the ability to accumulate and maintain an internal nitrogen reserve [11,38,39]. Hence, the holdfasts, which are larger than the propagules, may withstand warm and nutrient-poor conditions using stored nitrogen.

Ito et al. [22] showed that the number of regenerated shoots per holdfast length was higher for holdfasts that were fragmented into lengths of 1 mm than for the longer holdfasts (i.e., 2.5, 5, 10, and 20 mm in length) in *S. fusiforme*. Similarly, in the present study, the regeneration rate significantly increased with holdfast fragmentation from 5 mm to 1 mm in length. Ito et al. [22] also reported that the percentage of *S. fusiforme* holdfasts that geminated new shoots tended to decline in response to decreased temperature (from 23 °C to 17 °C) and irradiance (from 230 to 50 µmol photons m^{-2} s^{-1}). However, in the present study, the regeneration rates of new shoots from *S. fusiforme* holdfasts were not significantly affected by broader ranges of temperature (15 °C–30 °C and 30 °C–38 °C), nutrient availability, or irradiance, although the values tended to decrease in response to temperature elevation from 20 °C and 30 °C and nutrient enrichment. These results indicated that the regeneration of new shoots from holdfasts of this species is strongly regulated by physical stimulation (i.e., fragmentation) rather than abiotic environmental factors. This enhancement of vegetative reproduction by fragmentation is common in clonal plants and may be advantageous for reproduction in highly disturbed environments [2,3].

Loffler et al. [24] showed that the survival rate of *S. swartzii* was unaffected by the experimental removal (to mimic fish herbivory) of 50% of holdfast biomass but decreased when 75% of holdfast biomass was removed. In contrast, Ito et al. [22] and the present study showed that the filamentous holdfasts of *S. fusiforme* regenerated new shoots even after they were fragmented into 1-mm segments. Moreover, the present study showed that the regeneration rate was unaffected by reduced irradiance from 130 to 30 µmol photons m^{-2} s^{-1}, and the positive effect of fragmentation was not antagonized by the irradiance reduction. Hence, the holdfasts of this species can regenerate new shoots if the tiny segments remain in shaded refuges, such as crevices [34], to avoid intensive fish herbivory. However, holdfast growth decreased in response to reduced irradiance in the present study. Therefore, the survival rate of the holdfasts may depend on the light environment of these refuges.

Sargassum fusiforme is distributed from Hokkaido in northern Japan to Okinawa Prefecture in southern Japan [35]. The maximum seawater temperature and DIN concentration ranges during summer (between July and September) are 28.2 °C–29.1 °C and 1.1–4.5 µM, respectively, at several sites in Kagoshima Prefecture [26,27], near the southern distributional limit of this species. The results of the present study predict that *S. fusiforme* holdfasts can grow during the summer in Kagoshima Prefecture, and have the potential to survive under a further warming of 2 °C–3 °C [40], whereas their propagule survival during summer depends on the local nutrient environment in this region. The present study also showed that holdfast fragmentation enhanced vegetative reproduction. These traits may allow survival under the warm, nutrient-poor, and highly grazed environments in southern Japan, where local extinctions of other kelp and fucoid species have been reported [17]. However, further genetic approaches are required to quantify the contribution of vegetative and sexual reproduction to population persistence, and to determine the genotype that enables the growth of sexually-reproduced propagules into matured individuals with large holdfasts that exhibit vegetative reproduction.

Sargassum fusiforme is a popular food in China, Korea, and Japan, and therefore seeding methods for the cultivation of this species have been developed [22,41–43]. Ito et al. [22]

suggested that holdfasts harvested during spring can be stored by incubation at an irradiance of 1 μmol photons m^{-2} s^{-1} in order to suppress regeneration, and can be utilized for the seeds by means of a tank culture for 40 d after cutting them into segments <2.5 mm in length. This method seems to be more efficient and effective than the one based on sexual reproduction in southern Japan under warming, according to the results of the present study. However, little is known about differences in growth characteristics between regenerated shoots and sexually reproduced shoots. Further studies on the ecological and physiological traits of holdfasts and regenerated shoots of this species may provide insights for the conservation and restoration of marine macroalgal forests in southern Japan under ocean warming conditions, and for the improvement of seed production methods for this species.

4. Materials and Methods

4.1. Experiment 1: Combined Effects of Temperature and Nutrients on Propagules

Sargassum fusiforme propagules were cultured for 21 d in nine different treatments, consisting of three temperature levels (10 °C, 20 °C, and 30 °C) and three nutrient levels. These temperature levels were chosen based on the study of Baba [29], which showed that the growth rate of *S. fusiforme* propagules increased in response to elevated temperature from 10 °C to 25 °C, and was similar between 25 °C and 30 °C at an irradiance of 100 and 180 μmol photons m^{-2} s^{-1} in a 20-d experiment using nutrient-rich PESI [44] as the culture medium. The three nutrient levels were set as 10% PESI, 5% PESI, and SSW in order to evaluate the effect of reduced nutrient availability compared to natural levels. The DIN concentrations in the culture media were measured using an autoanalyzer with four replications.

In detail, matured shoots of *S. fusiforme*, in which many propagules were observed on the surface of female receptacles, were collected in June 2020 from a site in Yojiro (31°33′30″ N, 130°33′47″ E), Kagoshima Prefecture, southern Japan, and were transported to the laboratory in insulated cool boxes. These shoots were placed in a plastic container containing natural seawater for 7 d, and 54 propagules, which naturally dropped from the shoots to the container bottom, were collected using a Pasteur pipette. These propagules were placed in several petri dishes (9 mm in diameter) containing 30 mL of SSW, and were incubated at a temperature of 20 °C and an irradiance of 30 μmol photons m^{-2} s^{-1} with a 12 h light (L):12 h dark (D) photoperiod for 4 d until the start of the experiment.

The 54 propagules were assigned to one of the nine treatments. Each propagule was placed in six holes (one propagule per hole) of nine culture plates (P24F01S, AS ONE, Osaka, Japan) containing 2.5 mL of culture medium, and was incubated for 21 d at 130 μmol photons m^{-2} s^{-1} with a 12 h L:12 h D photoperiod. The culture medium was changed every 3 d. Photographs of these propagules were taken under a stereoscopic microscope using a digital camera before (initial value) and after culturing (final value), and the surface areas of all propagules were measured using Image J software [45]. The initial mean surface area (± standard deviation [SD]) was 0.096 ± 0.023 mm^{-2}. Relative growth rates (% d^{-1}) were calculated as 100 × ln (final value/initial value)/culture d.

All statistical analyses in the present study were performed using SPSS software version 20.0 (IBM, Armonk, NY, USA). The combined effects of temperature and nutrient availability on the relative growth rates were tested using a two-way ANOVA. When significant interactions between two factors were found, Tukey's multiple comparison tests were used to examine the differences among the nine treatments.

4.2. Experiment 2: Combined Effects of Temperature and Nutrients on Holdfasts

Holdfasts of *S. fusiforme* were cultured for 28 d in eight different treatments, consisting of four temperature levels (15 °C, 20 °C, 25 °C, and 30 °C) and two nutrient levels. The four temperature levels were chosen within the temperature range at the site of collection (ca. 15 °C during winter and ca. 30 °C during summer, H. Endo unpublished data) due to the lack of available information on the optimal holdfast growth temperature. The two nutrient levels were set at 25% PESI and SSW because the holdfast growth was very slow, and was slightly affected by nutrient enrichment using 5% and 10% PESI in our preliminary experiment. The DIN concentrations in the culture media were measured in the same manner as in experiment 1.

In detail, six *S. fusiforme* individuals with relatively large holdfasts were collected in May 2018 from the site in Yojiro, and 48 holdfast segments, 5 mm in length without shoots, were cut from the specimens (eight segments per individual). Each of the eight segments derived from an individual plant were placed in a petri dish (six dishes in total) and these dishes were incubated for 24 h at 20 °C and 130 µmol photons m^{-2} s^{-1} with a 12 h L:12 h D photoperiod.

At the beginning of the experiment, the wet weight of each segment (initial value) was measured using an electronic balance (0.1 mg accuracy) after the removal of excess moisture by blotting on paper towels. The initial mean wet weight (± SD) was 19.04 ± 6.49 mg. The 48 segments were randomized into eight groups of six specimens with a similar size distribution. The eight groups were each subjected to one of the eight different treatments. These specimens were placed in a petri dish (one segment per dish) containing 30 mL of culture medium and were maintained in incubators at 130 µmol photons m^{-2} s^{-1} with a 12 h L:12 h D photoperiod for 28 d. The culture medium in each dish was changed every 7 d. At the end of the experiment, the wet weight and number of newly regenerated shoots of the cultured segment (final value) were determined. Relative growth and regeneration rates (% d^{-1}) were calculated as 100 × ln (final value/initial value)/culture d. The combined effects of temperature and nutrients on the relative growth and regeneration rates were tested using two-way ANOVA and Tukey's multiple comparison tests.

4.3. Experiment 3: Effect of High Temperature on Holdfasts

Sargassum fusiforme holdfasts were cultured at five different temperatures (30 °C, 32 °C, 34 °C, 36 °C, and 38 °C). Six *S. fusiforme* individuals were collected in July 2018 and 30 holdfast segments, 5 mm in length, were cut from the specimens (five segments per individual). These segments were incubated for 24 h at 20 °C and 130 µmol photons m^{-2} s^{-1} with a 12 h L:12 h D photoperiod. The 30 segments were randomized into five groups of six specimens with a similar size distribution. The five groups were each subjected to one of the five temperatures. These specimens were placed in a petri dish (one segment per dish) containing 30 mL of SSW and were incubated at 130 µmol photons m^{-2} s^{-1} with a 12 h L:12 h D photoperiod for 28 d. The culture medium in each dish was changed every 7 d. The wet weight and shoot number of each segment were determined before and after culturing, and the relative growth and regeneration rates were calculated in the same manner as in experiment 2. The initial mean wet weight (± SD) was 16.11 ± 3.50 mg. The effect of temperature on relative growth and regeneration rates were tested using one-way ANOVA and Tukey's multiple comparison tests.

4.4. Experiment 4: Combined Effects of Fragmentation and Irradiance on Holdfasts

Holdfasts of *S. fusiforme* were cultured for 28 d in four different treatments, consisting of two fragmentation levels (cutting into five segments of 1 mm in length and a control without cutting) and two irradiance levels (30 and 130 µmol photons m^{-2} s^{-1}). The two fragmentation levels were selected based on a study by Ito et al. [22], which reported that the number of regenerated shoots per unit length of holdfasts was higher for the holdfasts that were fragmented into 1-mm lengths, than for the longer holdfasts (i.e., 2.5, 5, 10, and 20 mm in length). The two irradiance levels were chosen because the relative growth rate of the holdfasts was significantly affected by reduced irradiance from 130 to 30 µmol photons m^{-2} s^{-1} in our previous experiment.

Six *S. fusiforme* individuals were collected in May 2019, and 24 holdfast segments were cut (four segments per individual). These segments were incubated for 24 h at 20 °C and 130 µmol photons m^{-2} s^{-1} with a 12 h L:12 h D photoperiod. The 24 segments were randomized into four groups of six specimens with a similar size distribution. The four groups were each subjected to one of the four treatments. These specimens were placed in a petri dish (one segment per dish) containing 30 mL of SSW and were incubated under a 12 h L:12 h D photoperiod for 28 d. The culture medium in each dish was changed every 7 d. The wet weight and shoot number of each segment were measured before and after culturing and the relative growth and regeneration rates were calculated in the same manner as in experiments 2 and 3. The initial mean wet weight (± SD) was 11.52 ± 4.43 mg. The combined effects of fragmentation and irradiance on relative growth and regeneration rates were tested using two-way ANOVA.

Author Contributions: Conceptualization, H.E.; methodology, H.E., R.I. and S.O.; validation, H.E. formal analysis, H.E. and S.O.; investigation, T.S., Y.Y., Y.N. and H.M.; data curation, H.E.; writing—original draft preparation, H.E.; writing—review and editing, H.E., R.I. and S.O.; visualization, T.S. Y.Y., Y.N. and H.E.; supervision and project administration, H.E. All authors have read and agreed to the published version of the manuscript.

Funding: This work was partially supported by a JSPS KAKENHI Grant (Number 20K06187).

Acknowledgments: We thank R. Terada of Kagoshima University and T. Noda of Japan Fisheries Research and Education Agency for their helpful comments on the manuscript. We also thank M Matsuoka of Kagoshima University and S. Ninomiya of Kagoshima City Fisheries Cooperative for their assistance with sample collection.

Conflicts of Interest: All authors declare that the research was conducted in the absence of any commercial or financial relationships that could be construed as a potential conflict of interest.

References

1. Schmidt, A.; Schmid, M.W.; Grossniklaus, U. Plant germline formation: Common concepts and developmental flexibility in sexual and asexual reproduction. *Development* **2015**, *142*, 229–241. [CrossRef] [PubMed]
2. Roiloa, S.R.; Retuerto, R. Effects of fragmentation and seawater submergence on photochemical efficiency and growth in the clonal invader Carpobrotus edulis. *Flora* **2016**, *225*, 45–51. [CrossRef]
3. Zhou, J.; Li, H.L.; Alpert, P.; Zhang, M.X.; Yu, F.H. Fragmentation of the invasive, clonal plant Alternanthera philoxeroides decreases its growth but not its competitive effect. *Flora* **2017**, *228*, 17–23. [CrossRef]
4. Wang, J.; Xu, T.; Wang, Y.; Li, G.; Abdullah, I.; Zhong, Z.; Liu, J.; Zhu, W.; Wang, L.; Wang, D.; et al. A meta-analysis of effects of physiological integration in clonal plants under homogeneous vs. heterogeneous environments. *Funct. Ecol.* **2021**, *35*, 578–589. [CrossRef]
5. Dayton, P.K. Toward understanding of community resilience and the potential effects of enrichment to the benthos at McMurdo Sound, Antarctica. In *Proceedings of the Colloquium on Conservation Problems in Antarctica*; Parker, B.C., Ed.; Allen Press: Lawrence, KS, USA, 1972; pp. 81–95.
6. Coleman, M.A.; Wernberg, T. Forgotten underwater forests: The key role of fucoids on Australian temperate reefs. *Ecol. Evol.* **2017**, *7*, 8406–8418. [CrossRef]
7. Eger, A.M.; Marzinelli, E.; Gribben, P.; Johnson, C.R.; Layton, C.; Steinberg, P.D.; Wood, G.; Silliman, B.R.; Vergés, A. Playing to the positives: Using synergies to enhance kelp forest restoration. *Front. Mar. Sci.* **2020**, *7*, 544. [CrossRef]
8. Filbee-Dexter, K.; Wernberg, T. Substantial blue carbon in overlooked Australian kelp forests. *Sci. Rep.* **2020**, *10*, 1–6. [CrossRef] [PubMed]
9. Watanabe, K.; Yoshida, G.; Hori, M.; Umezawa, Y.; Moki, H.; Kuwae, T. Macroalgal metabolism and lateral carbon flows can create significant carbon sinks. *Biogeosciences* **2020**, *17*, 2425–2440. [CrossRef]
10. Smale, D.A. Impacts of ocean warming on kelp forest ecosystems. *New Phytol.* **2020**, *225*, 1447–1454. [CrossRef]
11. Gerard, V.A. The role of nitrogen nutrition in high-temperature tolerance of the kelp, *Laminaria saccharina* (Chromophyta). *J. Phycol.* **1997**, *33*, 800–810. [CrossRef]
12. Gao, X.; Endo, H.; Taniguchi, K.; Agatsuma, Y. Combined effects of seawater temperature and nutrient condition on growth and survival of juvenile sporophytes of the kelp *Undaria pinnatifida* (Laminariales; Phaeophyta) cultivated in northern Honshu, Japan. *J. Appl. Phycol.* **2013**, *25*, 269–275. [CrossRef]
13. Gao, X.; Endo, H.; Nagaki, M.; Agatsuma, Y. Growth and survival of juvenile sporophytes of the kelp *Ecklonia cava* in response to different nitrogen and temperature regimes. *Fish. Sci.* **2016**, *82*, 623–629. [CrossRef]
14. Gao, X.; Endo, H.; Nagaki, M.; Agatsuma, Y. Interactive effects of nutrient availability and temperature on growth and survival of different size classes of *Saccharina japonica* (Laminariales, Phaeophyceae). *Phycologia* **2017**, *56*, 253–260. [CrossRef]
15. Vergés, A.; Steinberg, P.D.; Hay, M.E.; Poore, A.G.; Campbell, A.H.; Ballesteros, E.; Heck, K.L.; Booth, D.J.; Coleman, M.A.; Feary, D.A.; et al. The tropicalization of temperate marine ecosystems: Climate-mediated changes in herbivory and community phase shifts. *Proc. R. Soc. B Biol. Sci.* **2014**, *281*, 20140846. [CrossRef] [PubMed]

16. Watanabe, Y.W.; Ishida, H.; Nakano, T.; Nagai, N. Spatiotemporal decreases of nutrients and chlorophyll-a in the surface mixed layer of the western North Pacific from 1971 to 2000. *J. Oceanogr.* **2005**, *61*, 1011–1016. [CrossRef]
17. Kumagai, N.H.; Molinos, J.G.; Yamano, H.; Takao, S.; Fujii, M.; Yamanaka, Y. Ocean currents and herbivory drive macroalgae-to-coral community shift under climate warming. *Proc. Natl. Acad. Sci. USA* **2018**, *115*, 8990–8995. [CrossRef]
18. Endo, H.; Nishigaki, T.; Yamamoto, K.; Takeno, K. Age-and size-based morphological comparison between the brown alga *Sargassum macrocarpum* (Heterokonta; Fucales) from different depths at an exposed coast in northern Kyoto, Japan. *J. Appl. Phycol.* **2013**, *25*, 1815–1822. [CrossRef]
19. Endo, H.; Nishigaki, T.; Yamamoto, K.; Takeno, K. Subtidal macroalgal succession and competition between the annual, *Sargassum horneri*, and the perennials, *Sargassum patens* and *Sargassum piluliferum*, on an artificial reef in Wakasa Bay, Japan. *Fish. Sci.* **2019**, *85*, 61–69. [CrossRef]
20. Yoshida, G.; Shimabukuro, H. Seasonal population dynamics of *Sargassum fusiforme* (Fucales, Phaeophyta), Suo-Oshima Is., Seto Inland Sea, Japan—development processes of a stand characterized by high density and productivity. *J. Appl. Phycol.* **2017**, *29*, 639–648. [CrossRef]
21. Low, J.K.; Fong, J.; Todd, P.A.; Chou, L.M.; Bauman, A.G. Seasonal variation of *Sargassum ilicifolium* (Phaeophyceae) growth on equatorial coral reefs. *J. Phycol.* **2019**, *55*, 289–296. [CrossRef]
22. Ito, R.; Terawaki, T.; Satuito, C.G.; Kitamura, H. Storage, cutting and culture conditions of filamentous roots of Hiziki *Sargassum fusiforme*. *Aquacult. Sci.* **2009**, *57*, 579–585, (In Japanese with English Abstract).
23. Yatsuya, K.; Kiyomoto, S.; Yoshida, G.; Yoshimura, T. Regeneration of the holdfast in 13 species of the genus *Sargassum* grown along the western coast of Kyushu, south-western Japan. *Jpn. J. Phycol.* **2012**, *60*, 41–45, (In Japanese with English Abstract).
24. Loffler, Z.; Graba-Landry, A.; Kidgell, J.T.; McClure, E.C.; Pratchett, M.S.; Hoey, A.S. Holdfasts of *Sargassum swartzii* are resistant to herbivory and resilient to damage. *Coral Reefs* **2018**, *37*, 1075–1084. [CrossRef]
25. Yamaguchi, A.; Furumitsu, K.; Yagishita, N.; Kume, G. Biology of herbivorous fish in the coastal areas of western Japan. In *Coastal Environmental and Ecosystem Issues of the East China Sea*; Ishimatsu, A., Lie, H.-J., Eds.; TERRAPUB and Nagasaki University: Nagasaki, Japan, 2010; pp. 181–2190.
26. Shimabukuro, H.; Terada, R.; Sotobayashi, J.; Nishihara, G.N.; Noro, T. Phenology of *Sargassum duplicatum* (Fucales, Phaeophyceae) from the southern coast of Satsuma Peninsula, Kagoshima, Japan. *Nippon Suisan Gakkaishi* **2007**, *73*, 454–460, (In Japanese with English Abstract). [CrossRef]
27. Tsuchiya, Y.; Sakaguchi, Y.; Terada, R. Phenology and enrivonmental characteristics of four *Sargassum* species (Fucales): *S. piluliferum*, *S. patens*, *S. crispifolium*, and *S. alternato-pinnatum* from Sakurajima, Kagoshima Bay, southern Japan. *Jpn. J. Phycol.* **2011**, *59*, 1–8, (In Japanese with English Abstract).
28. Nakashima, H.; Tanaka, T.; Yoshimitsu, S.; Terada, R. Phenology of three species of *Sargassum* (Fucales) and the long-term change of seaweed community structure from Kasasa, Kagoshima Prefecture, Japan. *Jpn. J. Phycol.* **2013**, *61*, 97–105, (In Japanese with English Abstract).
29. Baba, M. Effects of temperature and irradiance on germling growth in eight Sargassaceous species. *Rep. Mar. Ecol. Res. Inst.* **2007**, *10*, 9–20.
30. Graba-Landry, A.C.; Loffler, Z.; McClure, E.C.; Pratchett, M.S.; Hoey, A.S. Impaired growth and survival of tropical macroalgae (*Sargassum* spp.) at elevated temperatures. *Coral Reefs* **2020**, *39*, 475–486. [CrossRef]
31. Hay, K.B.; Millers, K.A.; Poore, A.G.; Lovelock, C.E. The use of near infrared reflectance spectrometry for characterization of brown algal tissue. *J. Phycol.* **2010**, *46*, 937–946. [CrossRef]
32. Endo, H.; Suehiro, K.; Kinoshita, J.; Gao, X.; Agatsuma, Y. Combined effects of temperature and nutrient availability on growth and phlorotannin concentration of the brown alga *Sargassum patens* (Fucales; Phaeophyceae). *Am. J. Plant Sci.* **2013**, *4*, 14–20. [CrossRef]
33. Li, J.; Liu, Y.; Liu, Y.; Wang, Q.; Gao, X.; Gong, Q. Effects of temperature and salinity on the growth and biochemical composition of the brown alga *Sargassum fusiforme* (Fucales, Phaeophyceae). *J. Appl. Phycol.* **2019**, *31*, 3061–3068. [CrossRef]
34. Loffler, Z.; Hoey, A.S. Microtopographic refuges enhance recruitment and survival but inhibit growth of propagules of the tropical macroalga *Sargassum swartzii*. *Mar. Ecol. Prog. Ser.* **2019**, *627*, 61–70. [CrossRef]
35. Yoshida, T. *Marine Algae of Japan*; Uchida Rokakuho: Tokyo, Japan, 1998. (In Japanese)
36. Hu, Z.M.; Li, J.J.; Sun, Z.M.; Gao, X.; Yao, J.T.; Choi, H.G.; Endo, H.; Duan, D.L. Hidden diversity and phylogeographic history provide conservation insights for the edible seaweed *Sargassum fusiforme* in the Northwest Pacific. *Evol. Appl.* **2017**, *10*, 366–378. [CrossRef]
37. Liu, L.; Lin, L. Effect of heat stress on *Sargassum fusiforme* leaf metabolome. *J. Plant Biol.* **2020**, *63*, 229–241. [CrossRef]
38. Gao, X.; Endo, H.; Taniguchi, K.; Agatsuma, Y. Genetic differentiation of high-temperature tolerance in the kelp *Undaria pinnatifida* sporophytes from geographically separated populations along the Pacific coast of Japan. *J. Appl. Phycol.* **2013**, *25*, 567–574. [CrossRef]
39. Endo, H.; Inomata, E.; Gao, X.; Kinoshita, J.; Sato, Y.; Agatsuma, Y. Heat stress promotes nitrogen accumulation in meristems via apical blade erosion in a brown macroalga with intercalary growth. *Front. Mar. Sci.* **2020**, *7*, 575721. [CrossRef]
40. IPCC. Climate change 2013: The physical science basis. In *Contribution of Working Group I to the Fifth Assessment Report of the Intergovernmental Panel on Climate Change (IPCC)*; Cambridge University Press: Cambridge, MA, USA, 2013.

41. Pang, S.J.; Chen, L.T.; Zhuang, D.G.; Fei, X.G.; Sun, J.Z. Cultivation of the brown alga *Hizikia fusiformis* (Harvey) Okamura: Enhanced seedling production in tumbled culture. *Aquaculture* **2005**, *245*, 321–329. [CrossRef]
42. Pang, S.J.; Zhang, Z.H.; Zhao, H.J.; Sun, J.Z. Cultivation of the brown alga *Hizikia fusiformis* (Harvey) Okamura: Stress resistance of artificially raised young seedlings revealed by chlorophyll fluorescence measurement. *J. Appl. Phycol.* **2007**, *19*, 557–565. [CrossRef]
43. Pang, S.J.; Shan, T.F.; Zhang, Z.H.; Sun, J.Z. Cultivation of the intertidal brown alga *Hizikia fusiformis* (Harvey) Okamura: Mass production of zygote-derived seedlings under commercial cultivation conditions, a case study experience. *Aquacult. Res.* **2008**, *39*, 1408–1415. [CrossRef]
44. Tatewaki, M. Formation of a crustaceous sporophyte with unilocular sporangia in *Scytosiphon lomentaria*. *Phycologia* **1966**, *6*, 62–66. [CrossRef]
45. Schneider, C.A.; Rasband, W.S.; Eliceiri, K.W. NIH Image to ImageJ: 25 years of image analysis. *Nat. Methods* **2012**, *9*, 671–675. [CrossRef] [PubMed]

 *plants*

MDPI

Article

Different Growth and Sporulation Responses to Temperature Gradient among Obligate Apomictic Strains of *Ulva prolifera*

Yoichi Sato [1,2,*], Yutaro Kinoshita [1,3], Miho Mogamiya [1], Eri Inomata [1], Masakazu Hoshino [4] and Masanori Hiraoka [3,*]

1 Bio-Resources Business Development Division, Riken Food Co., Ltd., Miyagi 985-0844, Japan; yuu_kinoshita@rikenfood.co.jp (Y.K.); mih_mogamiya@rikenfood.co.jp (M.M.); eri_inomata@rikenfood.co.jp (E.I.)
2 Nishina Center for Accelerator-Based Science, RIKEN, Saitama 351-0198, Japan
3 Usa Marine Biological Institute, Kochi University, Kochi 781-1164, Japan
4 Department of Algal Development and Evolution, Max Planck Institute for Developmental Biology, Max-Planck-Ring 5, 72076 Tübingen, Germany; mhoshino.sci@gmail.com
* Correspondence: yoi_sato@rikenfood.co.jp (Y.S.); mhiraoka@kochi-u.ac.jp (M.H.); Tel.: +81-22-395-4226 (Y.S.); +81-88-856-0426 (M.H.)

Abstract: The green macroalga *Ulva prolifera* has a number of variants, some of which are asexual (independent from sexual variants). Although it has been harvested for food, the yield is decreasing. To meet market demand, developing elite cultivars is required. The present study investigated the genetic stability of asexual variants, genotype (*hsp90* gene sequences) and phenotype variations across a temperature gradient (10–30 °C) in an apomictic population. Asexual variants were collected from six localities in Japan and were isolated as an unialgal strain. The *hsp90* gene sequences of six strains were different and each strain included multiple distinct alleles, suggesting that the strains were diploid and heterozygous. The responses of growth and sporulation versus temperature differed among strains. Differences in thermosensitivity among strains could be interpreted as the result of evolution and processes of adaptation to site-specific environmental conditions. Although carbon content did not differ among strains and cultivation temperatures, nitrogen content tended to increase at higher temperatures and there were differences among strains. A wide variety of asexual variants stably reproducing clonally would be advantageous in selecting elite cultivars for long-term cultivation. Using asexual variants as available resources for elite cultivars provides potential support for increasing the productivity of *U. prolifera*.

Keywords: macroalga; *Ulva prolifera*; obligate asexual strain; relative growth rate; sporulation; land-based cultivation; germling cluster method

Citation: Sato, Y.; Kinoshita, Y.; Mogamiya, M.; Inomata, E.; Hoshino, M.; Hiraoka, M. Different Growth and Sporulation Responses to Temperature Gradient among Obligate Apomictic Strains of *Ulva prolifera*. *Plants* **2021**, *10*, 2256. https://doi.org/10.3390/plants10112256

Academic Editor: Koji Mikami

Received: 29 September 2021
Accepted: 19 October 2021
Published: 22 October 2021

Publisher's Note: MDPI stays neutral with regard to jurisdictional claims in published maps and institutional affiliations.

Copyright: © 2021 by the authors. Licensee MDPI, Basel, Switzerland. This article is an open access article distributed under the terms and conditions of the Creative Commons Attribution (CC BY) license (https://creativecommons.org/licenses/by/4.0/).

1. Introduction

The green macroalga *Ulva prolifera* O.F. Müller, 1778 (Class Ulvophyceae) is an example of an alga showing isomorphic alternation of generations, with sporophytes and gametophytes that are morphologically indistinguishable. In the life cycle of *Ulva*, gametophytes of two mating types release biflagellate gametes with positive phototaxis, and the zygote develops into the sporophytic phase [1]. Sporophytes release quadriflagellate meiospores through meiosis, which develop into genetically separate gametophytes [2]. In addition to the sexual life cycle, several *Ulva* species, including *U. prolifera*, are known to have two types of obligate asexual life cycles without sexual reproduction via meiosis and conjugation, reproducing through biflagellate or quadriflagellate diploid zoids specialized for asexual development, these zoids have negative phototaxis [3]. These asexual zoids were termed "zoosporoids" [4,5]. The quadriflagellate zoosporoids of obligate asexual life history were a length of <10 µm; these were smaller in size than quadriflagellate meiospores of sexual life history (>11 µm length). On the other hand, biflagellate zoosporoids of obligate asexual life history were a length of 8–9 μm; these were distinguished from biflagellate games (6–7 μm length) [6]. These asexual variants are regarded as diploid thalli because the amount of DNA in the cells of asexual thalli is similar to that in the cells of the sporophytic thallus [6,7]. A recent study conducting genome and transcriptome analyses of *U. prolifera* suggests that the asexual thalli originally evolved via apomeiosis in sporophytic thalli [8] Another study has revealed that the two asexual variants of *U. prolifera* show high levels of heterozygosity in the *hsp90* gene, probably as a result of hybridization among genetically distinct gametophytes [9]. These findings indicate that evolution of the asexual variants of *U. prolifera* has been independently repeated from various genetically distinct sexual populations. Therefore, although asexual *U. prolifera* variants show obligate clonal reproduction, they may include a wide range of genotypes with various physiological characteristics.

The industrial use of *U. prolifera* has a long history, being harvested and used as food in Japan since the 10th century A.D. [10]. Since the 1990s, the artificial seedling method has been developed and the cultivation in estuaries and brackish water areas has been carried out, mainly around Shikoku Island, the 4th largest island of the Japanese Archipelago [11]. However, both cultivated and natural harvest yields have declined markedly due to various environmental factors, particularly salinity and temperature (e.g., [12]). To meet market demand, the "germling cluster" (GC) method was developed as a novel way to raise seedlings and land-based cultivation on an industrial scale began in Japan early this century [13]. Globally, the land-based cultivation of *Ulva* species is focused on biomass production in view of their rapid growth and lack of any requirements for freshwater resources or particular soils for cultivation [14–16]. However, it is important to develop a stable cultivation technique in order to achieve sustainable industrial production.

Developing an elite cultivar of *Ulva* species is an essential factor for successful land-based cultivation on a commercial scale, with, in particular, a requirement for a superior growth rate [17]. Since growth rates can vary substantially between cultivars of *Ulva*, both from within the same species [18–20] and among different *Ulva* species [21], it is important to compare and select from a wide variety of species and cultivars for optimal biomass production [16]. Asexual variants are particularly appropriate for the selection of elite cultivars because their genetic characteristics can be preserved even across many generations, which is very different from the F_1 hybridization techniques used with terrestrial crops. Understanding the growth characteristics of asexual variants with respect to environmental differences is also necessary in order to be able to select elite cultivars appropriate to each cultivation area. Additionally, *U. prolifera* thalli often produce and release zoids apically and become shorter in length at 20 °C and over [22]. Therefore, cultivars continuing vegetative growth without sporulation at 20 °C and over are appropriate for land-based cultivation. In short, to select elite cultivars it is important to verify the growth characteristics of different variants not only to optimize growth but also with regard to temperature effects on maturation. However, there have been few studies so far concerning the phenotypic differentiation of *U. prolifera* asexual variants.

The aim of the present study, therefore, is to identify the physiological characteristics for growth and sporulation responses of asexual variants of *U. prolifera* among recognized strains. Asexual thalli were collected from six localities in Japan and compared for their growth rate, carbon and nitrogen contents, and sporulation responses across a range of temperatures using the GC method for seedling production and cultivation to apply the industrial aquaculture perspective.

2. Results

2.1. Molecular Analysis

The *hsp90* gene sequences of the six strains showed the presence of alternative bases at some positions, indicating the presence of different alleles (Figure 1) and suggesting that these strains are heterozygous and diploid.

```
                 10         20         30         40         50         60         70         80         90        100
Strain 1  CTGGGGAATC GCGGAAGGCT GTTGAAAAYT CGCCATTCAT CGAAAAGYTG AAGAGGAAGG GKCTKGAGGT SCTGTTCATG GTTGATCCAA TTGATGAATA
Strain 2  .......... .......... ........C. .......... Y......T.. .......... .T..T..... T......... .......... ..........
Strain 3  .......... .......... ........C. .......... .......T.. .......... .T..T..... C......... .......... ..........
Strain 4  .......... .......... ........C. .......... .......T.. .......... .T..T..... Y......... .......... ..........
Strain 5                             ..C. .......... .......... .......... .T........ Y......... .......... ..........
Strain 6  ....R..... ......R... .......... .......... .......... .......... .......... .......... .......... ..........
                110        120        130        140        150        160        170        180        190        200
Strain 1  TGCCGTTCAG CARCTGAAGG ARTATGATGG YAAGAAGTTG GTCAGTGTGA CAAAGGAAGG TCTGGAGATT GAGGAAGATG ATGAGGAGAA GAAACGTAAG
Strain 2  .......... ..A....... .A........ C......... .......... .......... .......... .......... .......... ..........
Strain 3  .......... ..A....... .A........ C......... .......... .......... .......... .......... .......... ..........
Strain 4  .......... ..A....... .A........ C......... .......... .......... .......... .......... .......... ..........
Strain 5  .......... .......... .......... .......... ..Y....... .......R.. .........Y .......... .......... ..........
Strain 6  .......... .......... .......... M......... .......... .......... .......... .......... .......... ..........
                210        220        230        240        250        260        270        280        290        300
Strain 1  GAAGAAYTGA AGAGCAAGTT CGAAGAGTTG ACACGTGTCA TCAAAGACAT CCTTGCAGAC AAGGTCGAAA AAGTTGTYGT CTCAGACAGA ATCGTGGAYT
Strain 2  .......... .......... .......... .......... .......... .......... .......... .......C.. .......... ........T.
Strain 3  ......C... .......... .......... .......... .......... .......... .......... .......C.. .......... ........T.
Strain 4  ......C... .......... .......... .......... .......... .......... .......... .......C.. .......... ........T.
Strain 5  .......... .......... Y......... ..W.....S. .......... ...W...... .......... .......... ...R...... .....S....
Strain 6  .......... .......... .......... .......... .......... .......... .......... .......... ...M...... ..........
                310        320        330        340        350        360        370        380        390        400
Strain 1  CYCCATGTGT TCTGGTTACW GGTGAGTATG GYTGGAGTGC TAACATGGAG AGAATYATGA AGGCCCAAGC ACTTCGTGAT AGCAGCATGA GCTCGTACAT
Strain 2  .......... .........A .......... .C........ Y......... .....C.... ....T..... .......... .......... ....A.....
Strain 3  .C........ .........T .......... .C........ .......... .....C.... .......... .......... .......... ..........
Strain 4  .C........ .......... .......... .C........ .........R .....C.... .......... .......... .......... ....R.....
Strain 5  .......... Y......... .......... .......... Y......... .....C.... ....S..... .......... .......... ..........
Strain 6  .......... .........A .......... .......... .......... .......... .......... .......... .......... ..........
                410        420        430        440        450        460        470        480        490        500
Strain 1  GAGYTCAAAG AARACTCTYG AAATCAACCC AGAAAATGGC ATTGTCGAAG AGTTGCGGAA GAGAAGTGAG GCAGATAAGT CTGACAAGAC AGTTAAAGAC
Strain 2  ...T...... ..G.....C. .......... .......... ..C....... .......... .......... .......... .......... ..........
Strain 3  ...T...... ..G.....C. .......... .......... .......... .......... .......... .......... .......... ..........
Strain 4  ...T...... ..G.....C. .......... .......... ..Y....... .......... .......... .......... .......... ..........
Strain 5  .......... ..G....... .......... .......... .....Y.... .RY....... .......... .....Y.... ....Y..... ..........
Strain 6  .......... .......... .......... .......... .......... .......... ......Y... .......... .......... ..........
                510        520        530        540
Strain 1  TTRGTCCTTC TTTTGTTYGA GACTGCACTG CTGTCATCT
Strain 2  ..G....... .......T.. .......... .........
Strain 3  ..G..S.... .......T.. .......... .........
Strain 4  ..G....... .......T.. .......... .........
Strain 5  ..G....... .......T.. .......... .........
Strain 6  .......... .......... .......... .........
```

Figure 1. Comparison of *hsp90* gene sequences from *Ulva prolifera* thalli (undergoing an asexual life cycle) collected from six Japanese strains (see Table 1). Dots indicate identity with Strain 1; blanks indicate deletions; and double rows of dots and the following characters indicate the presence of different alleles: K, either G or T; M, either A or C; R, either A or G; S, either C or G; W, either A or T; Y, either C or T.

Table 1. Basic information on the *Ulva prolifera* strains collected from six estuarine localities in Japan for use in this study.

River	City or Town, Prefecture or Subprefecture	Strain No.	Latitude and Longitude
Oboro	Akkeshi, Kushiro	1	43°04′33.7″ N, 144°50′16.6″ E
Sekiguchi	Yamada, Iwate	2	39°28′28.5″ N, 141°57′03.3″ E
Orikasa	Yamada, Iwate	3	39°26′57.4″ N, 141°57′43.5″ E
Sakari	Ofunato, Iwate	4	39°04′50.0″ N, 141°43′10.0″ E
Natori	Natori, Miyagi	5	38°10′49.6″ N, 140°56′51.7″ E
Takeshima	Shimanto, Kochi	6	32°57′44.5″ N, 132°58′34.0″ E

The Oboro River is on Hokkaido Island, the Takeshima River is on Shikoku Island, and the remaining rivers are in northeastern Honshu.

2.2. *Growth Rate and Sporulation at Different Temperatures*

The relative growth rate (RGR) at different cultivation temperatures differed among strains. The mean RGR of Strain 1 varied narrowly over the range 0.3–0.4 and was not significantly influenced by temperature (p = 0.298, Figure 2, Strain 1), and the values of the maximum were about 1.3 times those of the minimum. However, those of other localities varied among temperature and there were significantly differences by post-hoc tests ($p < 0.01$, Figure 2, Strain 2–6). Although no clear peaks of RGR of Strain 2 and 4–6 were detected, maximum values were detected at 20–30 °C (Figure 2, Strains 2 and 4–6). The maximum values were 2.2–2.9 times those of the minimum. However, RGRs for Strain 3 indicated a clear peak with a mean of 0.55 at 20 °C, which is 3 times faster than the value at 10 °C (Figure 2). The RGRs of Strains 1 and 3 were significantly higher than those of other strains at 10 °C ($p < 0.05$) and 20 °C ($p < 0.01$), respectively (Supplemental Table S1). Throughout the cultivation period, no sporulating cells were detected in any of the thalli incubated at 10 or 15 °C (Figure 3). However, from the 2nd day of cultures at 20 °C or above, sporulating cells were already present in Strain 4 thalli (Figure 3), and in the thalli of Strains 1–3, and 5 from the 4th day of culture (Figure 3). In Strains 3–5, sporulating cells occupied more than half the total thallus area at 30 °C (data not shown). Strain 6 thalli showed no evidence of sporulation in cultures incubated below 25 °C, sporulation only began from the 8th day of culture (Figure 3) and was limited to only the tip of the thallus (data not shown).

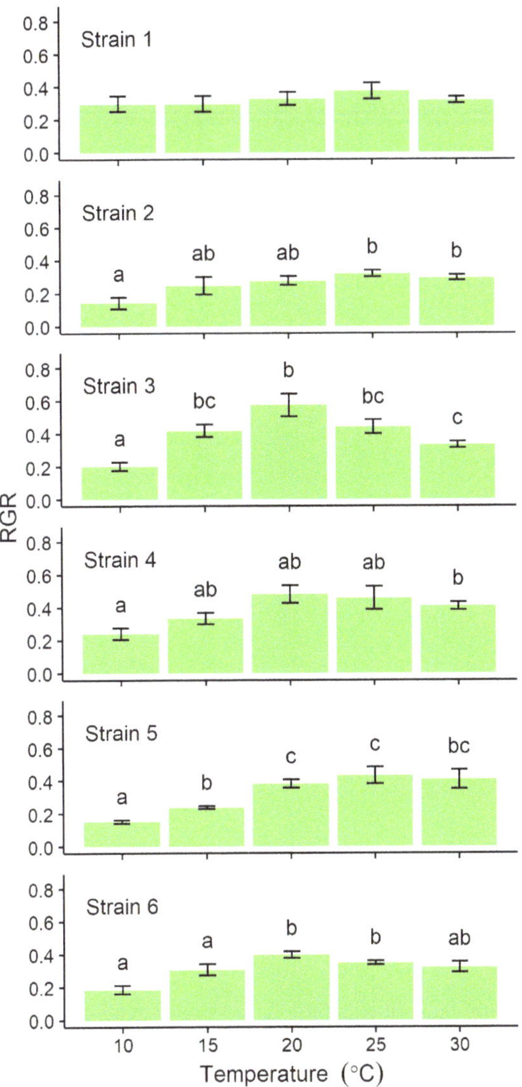

Figure 2. Relative growth rate (RGR) of *Ulva prolifera* thalli from each of six localities incubated in vitro at one of five different temperatures. The RGRs ($n = 5$) were calculated from four consecutive samples linearly arranged between 0.01 and 0.1 g (Supplemental Figure S1). Error bars indicate standard error of the mean. Different lowercase letters indicate significant differences ($p < 0.05$) among different temperatures. The results of statistical analysis among strains at each cultivation temperature were shown in Supplemental Table S1.

2.3. Carbon and Nitrogen Content at Different Temperatures

The carbon content of thalli for all six strains ranged between 0.337 ± 0.005 and 0.396 ± 0.001 mg mg^{-1}, with no obvious peak at any particular incubation temperature, although differences detected among incubation temperatures were significant for Strains 1–3 (Figure 4).

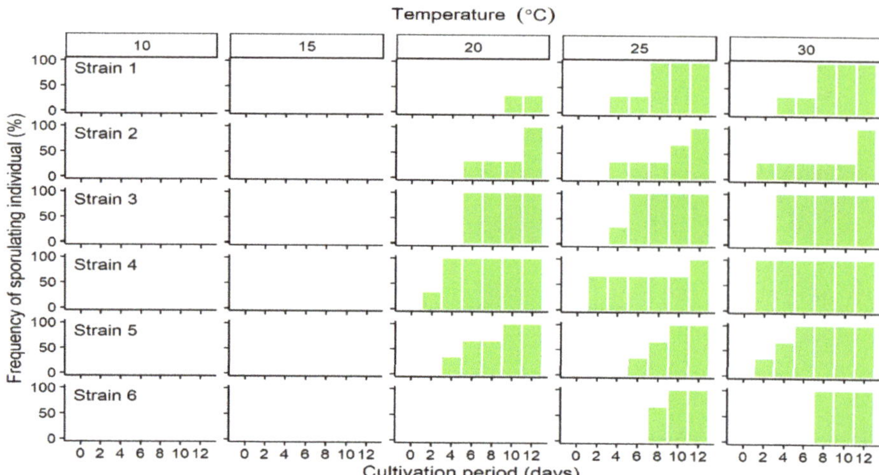

Figure 3. Changes in the frequency of occurrence of sporulating individuals of *Ulva prolifera* incubated in vitro at one of five different temperatures. Values are means ± standard error; $n = 5$ individuals.

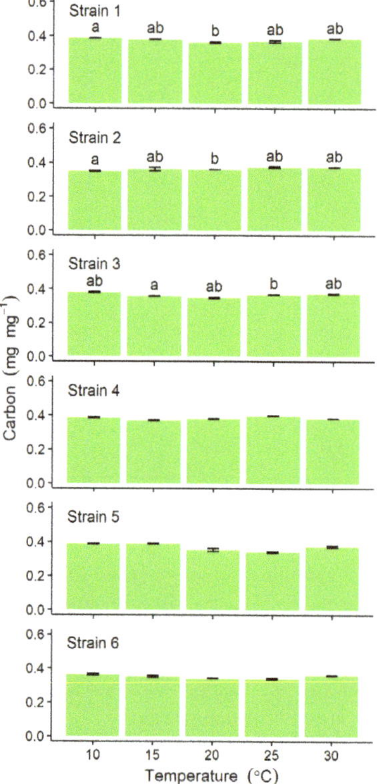

Figure 4. Comparison of carbon content in six strains of *Ulva prolifera* incubated in vitro at one of five different temperatures. Values are means ± standard error; $n = 5$ individuals. Different lower-case letters indicate significant differences ($p < 0.05$) among different temperatures. The results of statistical analysis among strains at each cultivation temperature were shown in Supplemental Table S1.

Nitrogen content ranged from 0.035 ± 0.001 (at 10 °C) to 0.055 ± 0.001 mg mg^{-1} (at 30 °C), tending to increase with increasing temperature, for all except Strain 6 (Figure 5). In the latter, no significant differences were detected among different temperature incubations, the values ranging between 0.037 ± 0.001 and 0.0433 ± 0.002 mg mg^{-1} (Figure 5).

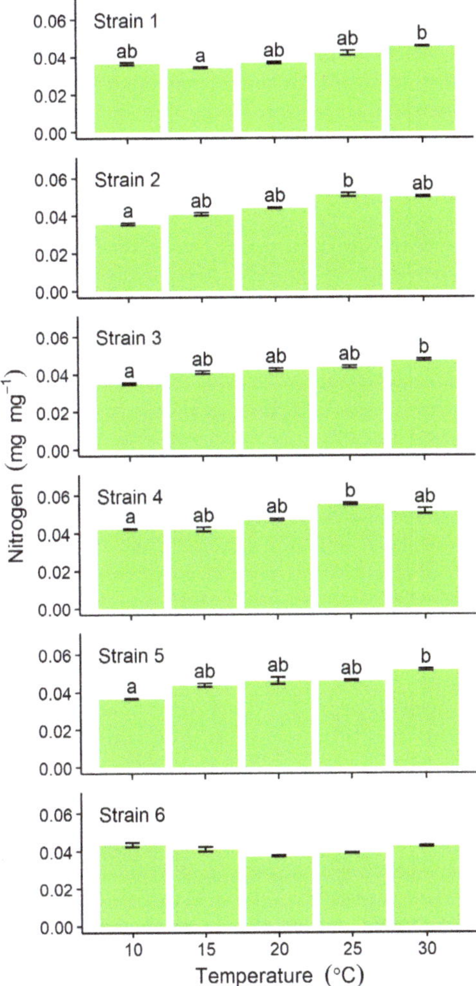

Figure 5. Comparison of nitrogen content in six strains of *Ulva prolifera* incubated in vitro at one of five different temperatures. Values are means ± standard error; $n = 5$ individuals. Different lowercase letters indicate significant differences ($p < 0.05$) among different temperatures. The results of statistical analysis among strains at each cultivation temperature were shown in Supplemental Table S1.

3. Discussion

The optimum temperature for growth of *U. prolifera* is known to vary according to sampling locality (strain) within the range 15–25 °C (reviewed by [23]), and the thalli generally mature and release zoids at 20 °C or higher [22]. However, previous studies did not identify the generation or type of life cycle of the specimens used in culture experiments. The present study revealed that different strains of asexual thalli of *U.*

prolifera had differences in thermosensitivity of growth and sporulation, and that this may be connected with the presence of different *hsp90* genotypes among these different strains. The growth rate temperature optima range within 20–25 °C, which is similar to that reported in previous studies [23]. It is noteworthy that for most strains the growth rate showed significant thermosensitivity. However, the growth rate of Strain 1 was not significantly influenced by temperature, suggesting that it can maintain growth even at lower temperatures. Strain 1 is from the northern Pacific coast of Hokkaido, which is close to the northern limit of distribution of *U. prolifera* [24], so it might be expected to have a lower temperature tolerance than other strains. With regard to sporulation, only Strain 6 thalli did not sporulate at 20 °C within 12 days, demonstrating a clearly different thermosensitivity from other strains, with a greater tolerance to high temperatures for maintaining vegetative growth.

It might be expected that the differences observed in growth rates and temperature tolerance among strains are connected with the environmental characteristics of the locality from which they were obtained. However, the results of the present study indicate that strains from neighboring localities have clearly different thermosensitivity; for instance, the localities of Strains 2 and 3 are only 5 km apart. Considering the genotypic differences among strains of *hsp90*, the phenotypic differences would be caused by genetic background. This suggests that the characteristics of each strain are not necessarily closely adapted to the environmental conditions at the locality in which it is found, and it is considered that the phenotypic diversity of this alga is itself high and not dependent upon site-specific environmental conditions. Hiraoka and Higa (2016) [3] proposed that *U. prolifera* had evolved from an ancestral marine species to become a true estuarine species: firstly, the sexual population adapted to low salinity conditions, and then a number of different asexual generations arose from genetically variable sexual ancestors, with natural selection finally producing an array of specialized asexual thallus genotypes that efficiently occupy the estuarine habitat. The variable thermosensitivity of asexual variants among localities could be interpreted as the result of the evolution and adaptation processes of this alga. The variation of *hsp90* genotypes observed among the six strains may be a manifestation of the phenotypic differentiation among them. Distinguishing among the genotypes affecting the phenotypes requires further study.

Measurements of the net photosynthetic rate and RGRs of *U. prolifera* collected from green tides in China have peak at 18–22 °C with a marked decline at 26 °C [25]. However, in the present study, carbon content was not observed to vary across different culture temperatures for all strains. This demonstrates the potential for stable carbon fixation among strains of *U. prolifera* regardless of temperature fluctuations, implying also a potential for CO_2 mitigation by *U. prolifera* which could be calculated from yield data.

Nitrogen content, however, varied with temperature for all strains except Strain 6, despite the presence of nutrients sufficient for culture conditions. In previous studies, the nitrogen content of *U. prolifera* collected from eutrophic areas of the Yellow Sea, was reported to be 3.6% [26]. However, values of 3–4% were found in wild-collected thalli from the same locality as Strain 6, where the dissolved inorganic nitrogen concentration measured was found to be insufficient for optimum growth [27]. In the present study, nitrogen content was in the range 2.8–5.1% across different incubation temperatures, suggesting that the assimilation capacity for nitrogen is influenced by temperature. Raven and Geider (1988) previously reported that temperature influences the nutrient-uptake rates via Q_{10} effects on algal metabolism [28]. The nitrogen content might therefore reflect the physiological response to differences in temperature which was a variable among the strains in the present study.

Many green algae show rapid nutrient uptake rates, contributing to the removal of excess nutrients in the water column [29,30]. According to the results of the present study, nutrient uptake kinetics might differ among strains and this may be useful for optimizing temperature-dependent quantitative removal of nitrogen from water column in land-based cultivation. This suggests a clear future requirement to ascertain the nutrient-uptake kinetics of each strain of asexual variant.

For practicing land-based cultivation on an industrial scale, it is impractical (and uneconomic) to have to adjust the seawater temperature in the tank by external means. Therefore, seawater pumped from offshore or from saline wells, with seasonally fluctuating temperatures, needs to be used as it is for land-based cultivation. In order to improve and maximize productivity for such seasonal changes in seawater temperature, information about growth and sporulation responses of asexual variants is required, as in the present study.

Currently, several land-based cultivation facilities in southern Japan are facing decreases in productivity due to reproductive maturation and pausing of growth at 20 °C or higher in summer. The use of cultivars with where sporulation does not occur until much higher temperatures, such as Strain 6 in the present study, would be one means to enable stable year-round cultivation in these southern areas. In contrast, higher productivity in the low winter temperature of northern areas require strains with higher growth rates at such temperatures. From the results of the present study, the growth rates of Strains 1 and 3 were significantly higher than other strains at 10 °C and 20 °C, respectively, so these strains can be regarded as elite cultivars at those temperatures.

It might also be effective to use these cultivars seasonally according to observed changes in seawater temperature. The present study revealed that asexual variants of *U. prolifera* cover a wide phenotypic range of thermosensitivity as a result of natural selection. With regard to preserving valuable characteristics as an elite cultivar, selection from asexual variants is a useful technique, because in sexual strains the occurrence of recombination may result in the loss of the required optimal responses. Therefore, evaluating and utilizing of these asexual variants as a resource pool of candidates for elite cultivars will help to support optimization of productivity and expand the cultivable area of algae such as *U. prolifera*.

4. Materials and Methods

4.1. Collection and Stock Maintenance of Thalli

Ulva prolifera thalli were collected from the estuaries of six Japanese rivers (see Table 1). To confirm the life cycle of all thalli collected at each locality, sporulation and releasing zoids were conducted according to Hiraoka et al. (2003) [6]. Zoids of samples from sites 1 to 4 and 6 were found to be biflagellate. These thalli were confirmed as asexual variants, since their zoids showed negative phototaxis and were obviously bigger than both male and female gametes reported in previous studies [6]. Zoids of site 5 were quadriflagellate. Thalli cultured from the quadriflagellate site 5 zoids released the same type of quadriflagellate zoids again. More than two generations were repeated and all released quadriflagellate zoids, confirming that site 5 thalli were obligate asexual variants. A unialgal culture strain was established for each locality (Table 1, Strains 1–6) at Usa Marine Biological Institute, Kochi University. All strains were transported to the Yuriage Factory, Riken Food Co., Ltd., in Natori City, Miyagi Prefecture, and their seeding stocks for the growth and maturation experiments were prepared according to the GC method for unattached (free-floating) macroalgal culture [13]. Thallus clusters were produced according to the method of Hiraoka et al. (2020) with slight modifications [21]. Synchronous zoid formation in each strain was induced by cutting a well-developed thallus into small fragments of 1–2 mm in length, which were washed in sterilized fresh water for approximately 10 s and cultured in a Petri dish containing 40 mL Enriched Seawater (ES) medium [31] at 20 °C under a 12 h:12 h L:D cycle, with light of 150 µmol photons m^{-2} s^{-1}. Under these conditions, thallus fragments released biflagellate (Strains 1–4 and 6) or quadriflagellate (Strain 5) zoids within 3 days.

Aliquots of zoid suspension densely concentrated using their phototactic response were placed in Petri dishes, adjusted to a density of >10^4 zoids per 1 mL medium, and incubated under the same conditions as above. After 3 weeks, germlings grew at a high density on the bottom of the dish and attached to one another to form aggregations with the appearance of a green mat. The aggregations were scraped off the dish without harming them, torn into numerous small clusters and cultured with gentle aeration, allowing to drift freely within the vessel. When they attained a length of 1 mm or more, they were statically stocked under weak light (<50 µmol photons m^{-2} s^{-1} under 12 h:12 h L:D cycle) at 20 °C for one week until used for growth-rate and sporulation-response experiments.

4.2. Molecular Analysis

Total genomic DNA was extracted from a small fragment of living material using BT Chelex® 100 Resin (Bio-Rad Cat# 143-2832, Hercules, CA, USA). A fragment of approximately 1 cm of each sample was ground in a 2 mL tube with 100 µL of 10% Chelex solution using a homogenization pestle at room temperature and incubated at 95 °C for 20 min, shaken at the middle and end of the 20 min period. The mixture was then cooled and centrifuged at 4000 rpm for 2 min.

Part of the sixth exon of the *hsp90* gene sequence was amplified using the primer pair of *hsp90*-6F (5′-GCAGACCCAGAAAGTGATCTATTAYATCA-3′) and *hsp90*-6R (5′-GCAGGYTCATCCAGACTAAATCC-3′), as reported by Ogawa et al. (2014) [9]. PCR amplifications were carried out using KOD FX Neo (ToYoBo, Osaka, Japan) and performed using a thermal cycler for 35 cycles of denaturation at 98 °C for 10 s, annealing at 55 °C for 30 s, and extension at 68 °C for 30 s, followed by a final hold at 10 °C. PCR products were sequenced by Fasmac (Atsugi, Kanagawa, Japan).

4.3. Growth Rate at Different Temperatures

To reduce the lag phase growth of the stocked materials, hundreds of thallus clusters for each strain were pre-cultured in a round 3L-flask with continuous aeration for 7 days. The flask was filled with sterilized seawater containing half-strength ES medium [31]. Temperature and light conditions for growth of germlings were as described above. The medium was changed every day. When the thallus clusters grew to 5–10 mm in length in this pre-culture, 8–12 clusters (0.01 g-wet, Figure 6a) were transferred to 500 mL-flasks and cultured with aeration at 10, 15, 20, 25, 30 °C for 8 days (Figure 6b). Light was provided from an LED unit (3LH-64, NK System Co., Ltd. Osaka, Japan) at 150 µmol photons $m^{-2}s^{-1}$ with a 12 h:12 h L:D cycle in the incubator (CN-40A, Mitsubishi Electric Engineering Co., Ltd., Tokyo, Japan). Half-strength ES medium was used as culture medium and changed every day. The temperature range and light intensity were set according to the previous study about the RGR of this alga vs. abiotic conditions [21], the RGR was saturated at a light intensity of >67 µmol photons m^{-2} s^{-1} and indicated broad ranges from 10 °C to 30 °C. The pre-culture experiment decided the enrichment medium condition; we confirmed that the half-strength of ES medium could be sufficient for the RGR by changing it daily. To determine the fresh mass of living material without causing damage by drying, thallus clusters were held between sterilized paper towels four times to carefully remove water on the surface, immediately placed in a Petri dish (9 cm in diameter) filled with half-strength ES medium on an electronic balance (0.1 mg accuracy), quantified, and returned to the same culture condition. This mass measurement was made within a few minutes at the end of the light period every day, equally spaced at 24 h intervals. Relative growth rate (RGR: the continuously accelerating growth of algae during the exponential phase; Supplemental Figure S1) was calculated using the following equation:

$$RGR = (\ln W_1 - \ln W_0)\ day^{-1}$$

where W_0 is the initial fresh mass in the culture at time zero, and W_1 is the mass after 24 h.

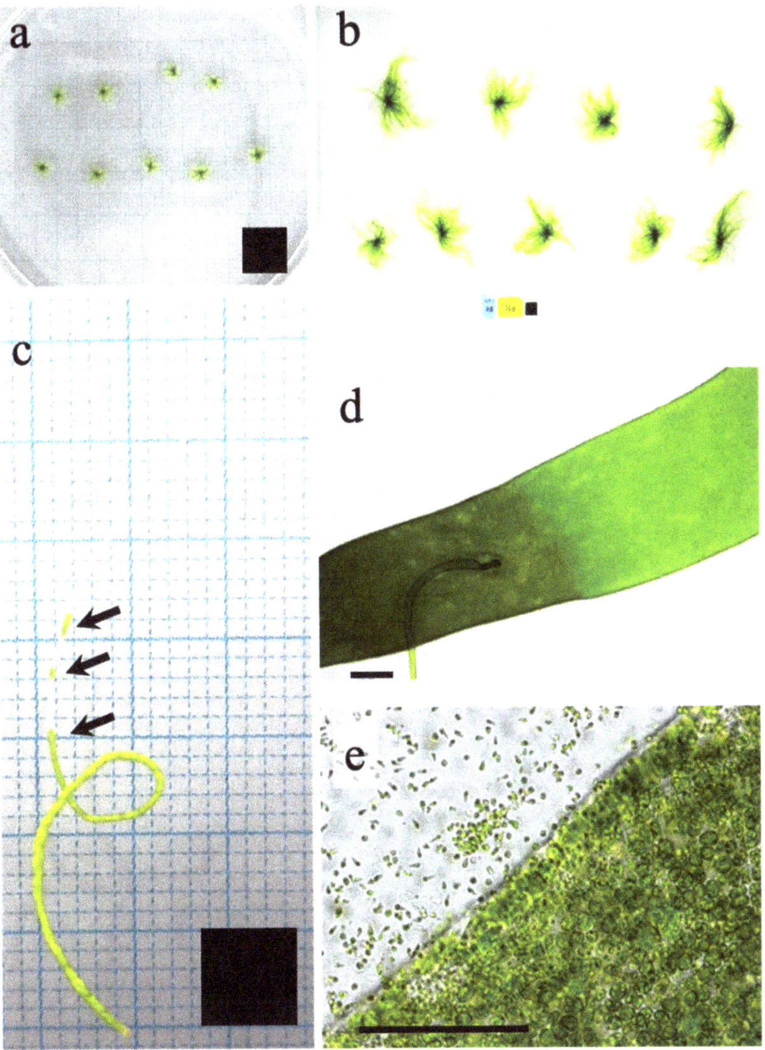

Figure 6. Representative images of *Ulva prolifera* used for the present study. (**a**,**b**) Thalli produced by the germling cluster method for the growth experiment: (**a**) initial thalli; (**b**) thalli after 8 d of cultivation. (**c**–**e**) Thallus for sporulation experiment: (**c**) sporulating individual with sporulated cells (arrows), (**d**) dark coloration indicating formation of sporulated cells, and (**e**) zoids released from sporulating part of thallus. Solid squares in (**a**–**c**) indicate 1 cm^2; bars (**d**,**e**), 300 μm.

4.4. Sporulation at Different Temperatures

Between five and eight precultured thallus clusters were selected and separated individually in the center part of clusters being careful not to injure the thallus and interfere with the release the sporulation inhibitor [32]. Three intact thalli (length 1 mm) from each strain were selected and each individual placed in a separate 500 mL flask and cultured at 10, 15, 20, 25, 30 °C under 150 μmol photons $m^{-2}s^{-1}$ with a 12 h:12 h L:D cycle for 12 days, using the same incubators and LED units as for the growth rate experiments. All were incubated in half-strength ES medium renewed every 2 days. When the medium was changed, the thalli were placed in a 9 cm Petri dish filled with half-strength ES medium and the thallus surface was observed by light microscopy for the presence or absence of sporulation (Figure 6c–e) and, using a digital camera attachment, recorded as digital images.

4.5. Carbon and Nitrogen Contents at Different Temperatures

For all strains used in the growth rate experiment at different temperatures, five clusters were randomly selected after final measurements had been taken. Since the light intensity used was above the compensation irradiance for photosynthesis [23], and the culture medium (half-strength ES; approximately 420 μM as nitrate) was changed every day, the thalli were considered to be supplied with sufficient carbon and nitrogen for normal growth to occur. Seawater was carefully blotted from the thallus surface of the cluster samples, which were placed in a dry oven (EYELA WFO-500, Tokyo Rikakikai Co., Ltd., Tokyo, Japan) for 12 h at 90 °C. The dried thalli were then each pulverized with a pestle and mortar and carbon and nitrogen content were measured using a CHN analyzer (Flash 2000, Thermofisher Scientific, Waltham, MA, USA).

4.6. Statistical Analysis

All data are presented as mean ± S.E. Significant differences in RGR, carbon content, and nitrogen content among different cultivation temperatures and different strains were identified by the Kruskal–Wallis test followed by Steel–Dwass multiple comparison tests. A nonparametric procedure was chosen because not all of the data were normally distributed or homoscedastic.

Supplementary Materials: The following are available online at https://www.mdpi.com/article/10.3390/plants10112256/s1, Figure S1: Upper figure: Wet weight variations in thallus clusters of six strains of *Ulva prolifera* incubated in vitro at one of five different temperatures, 150 μmol photons m^{-2} s^{-1} (12/12 h light/dark cycle) in half-strength ES medium with aeration for 8 d. Lower figure: the same data expressed as natural logarithms. The slopes for each combination of strain and temperature were used to generate the relative growth rate values in Figure 2. Table S1: Differences in RGR, Carbon, and nitrogen contents among five strains of *Ulva prolifera* at 10, 15, 20, 25, and 30 °C. Data were analyzed using Kruskal–Wallis test followed by post-hoc Scheffe's test for multiple comparisons.

Author Contributions: Conceptualization, Y.S.; methodology, Y.K., M.M., E.I., M.H. (Masakazu Hoshino); validation, Y.S.; formal analysis, Y.S. and E.I.; investigation, Y.S., Y.K., M.M., M.H. (Masakazu Hoshino). and M.H. (Masanori Hiraoka); data curation, Y.S. and M.H. (Masanori Hiraoka); writing—original draft preparation, Y.S.; writing—review and editing, M.H. (Masakazu Hoshino), and M.H. (Masanori Hiraoka); visualization, E.I. and Y.S.; supervision and project administration, Y.S. and M.H. (Masanori Hiraoka). All authors have read and agreed to the published version of the manuscript.

Funding: This work partially supported by basic-science research funding from Riken Food Co., Ltd., in 2019–2021 to Y.S., M.M., and Y.K.; and from the JST-OPERA Program (Grant Number JPMJOP1832) to Y.S., Y.K., and M.H. (Masanori Hiraoka).

Institutional Review Board Statement: Not applicable.

Informed Consent Statement: Not applicable.

Data Availability Statement: Data is contained within the article and supplementary material.

Acknowledgments: We thank H. Nagoe, M. Hoshi, and Y. Chiyokawa at the Yuriage Factory of Riken Food Co., Ltd., for their support with statistical analysis and cultivation studies. We also thank A. Kokubun and T. Taki of Riken Vitamin Co., Ltd., for their support with analysis of carbon and nitrogen contents.

Conflicts of Interest: All authors declare that the research was conducted in the absence of any commercial or financial relationships that could be construed as a potential conflict of interest.

References

1. Fjeld, A.; Lovlie, A. Genetics of multicellular marine algae. In *The Genetics of Algae*; Lewin, R.A., Ed.; Blackwell Scientific Publications: Oxford, UK, 1976; pp. 219–235.
2. van den Hoek, C.; Mann, D.G.; Jahns, H.M. *Algae. An Introduction to Phycology*; Cambridge University Press: Cambridge, UK, 1995; pp. 391–408.
3. Hiraoka, M.; Higa, M. Novel distribution pattern between coexisting sexual and obligate asexual variants of the true estuarine macroalga *Ulva prolifera*. *Ecol. Evol.* **2016**, *6*, 3658–3671. [CrossRef]
4. Bliding, C. Über. Enteromorpha intestinalis und compressa. *Bot. Not.* **1948**, *2*, 123–136.
5. Bliding, C. A critical survey of European taxa in Ulvales. Part, I. Capsosiphon, Percursaria, Blidingia, Enteromorpha. *Opera Botan.* **1963**, *8*, 1–160.
6. Hiraoka, M.; Dan, A.; Shimada, S.; Hagihira, M.; Migita, M.; Ohno, M. Different life histories of *Enteromorpha prolifera* (Ulvales, Chlorophyta) from four rivers on Shikoku Island, Japan. *Phycologia* **2003**, *42*, 275–284. [CrossRef]
7. Hiraoka, M.; Shimada, S.; Ohno, M.; Serisawa, Y. Asexual life history by quadriflagellate swarmers of *Ulva spinulosa* (Ulvales, Ulvophyceae). *Phycol. Res.* **2003**, *51*, 29–34. [CrossRef]
8. Ichihara, K.; Yamazaki, T.; Miyamura, S.; Hiraoka, M.; Kawano, S. Asexual thalli originated from sporophytic thalli via apomeiosis in the green seaweed Ulva. *Sci. Rep.* **2019**, *9*, 13523. [CrossRef] [PubMed]
9. Ogawa, T.; Ohki, K.; Kamiya, M. High heterozygosity and phenotypic variation of zoids in apomictic *Ulva prolifera* (Ulvophyceae) from brackish environments. *Aquat. Bot.* **2014**, *120*, 185–192. [CrossRef]
10. Miyashita, A. *Marine Algae as Food*; Hosei Daigaku Shuppankyoku: Tokyo, Japan, 1974; p. 315. Available online: http://pi.lib.uchicago.edu/1001/cat/bib/5151556 (accessed on 15 September 2021).
11. Dan, A.; Ohno, M.; Matsuoka, M. Cultivation of the green alga *Enteromorpha prolifera* using chopped tissue for artificial seedling. *Aquac. Sci.* **1997**, *45*, 5–8, (In Japanese with English Abstract).
12. Pandey, R.S.; Ohno, M. An ecological study of cultivated *Enteromorpha*. *Usa Mar. Biol. Inst. Kochi Univ.* **1985**, *7*, 21–31.
13. Hiraoka, M.; Oka, N. Tank cultivation of *Ulva prolifera* in deep seawater using a new "germling cluster" method. *J. Appl. Phycol.* **2008**, *20*, 97–102. [CrossRef]
14. Bird, M.I.; Wurster, C.M.; de Paula Silva, P.H.; Bass, A.M.; de Nys, R. Algal biochar–production and properties. *Biores. Tech.* **2011**, *102*, 1886–1891. [CrossRef]
15. Mata, L.; Magnusson, M.; Paul, N.A.; de Nys, R. The intensive land-based production of the green seaweeds *Derbesia tenuissima* and Ulva ohnoi: Biomass and bioproducts. *J. Appl. Phycol.* **2016**, *28*, 365–375. [CrossRef]
16. Lawton, R.J.; Sutherland, J.E.; Glasson, C.R.K.; Magnusson, M.E. Selection of temperate Ulva species and cultivars for land-based cultivation and biomass applications. *Algal. Res.* **2021**, *56*, 102320. [CrossRef]
17. Sato, Y. Cultivation of seaweed in land based tank system in Japan. *Agric. Biotechnol.* **2021**, *5*, 44–47. (In Japanese)
18. Lawton, R.J.; Mata, L.; de Nys, R.; Paul, N.A. Algal bioremediation of waste waters from land-based aquaculture using Ulva: Selecting target species and strains. *PLoS ONE* **2013**, *8*, e77344. [CrossRef] [PubMed]
19. Fort, A.; Mannion, C.; Fariñas-Franco, K.M.; Sulpice, R. Green tides select for fast expanding Ulva strains. *Sci. Total Environ.* **2020**, *698*, 134337. [CrossRef]
20. Fort, A.; Lebrault, M.; Allaire, M.; Esteves-Ferreira, A.A.; McHale, M.; Lopez, F.; Fariñas-Franco, J.M.; Alseekh, S.; Fernie, A.R.; Sulpice, R. Extensive variations in diurnal growth patterns and metabolism among Ulva spp. Strains. *Plant Physiol.* **2019**, *180*, 109–122. [CrossRef]
21. Hiraoka, M.; Kinoshita, Y.; Higa, M.; Tsubaki, S.; Monotilla, A.P.; Onda, A.; Dan, A. Fourfold daily growth rate in multicellular marine alga *Ulva meridionalis*. *Sci. Rep.* **2020**, *10*, 12606. [CrossRef] [PubMed]
22. Hiraoka, M.; Dan, A.; Hagihira, M.; Ohno, M. Growth and maturity of clonal thalli in *Enteromorpha prolifera* under different temperature conditions. *Nippon Suisan Gakkaishi* **1999**, *62*, 302–303. (In Japanese) [CrossRef]
23. Xiao, J.; Zhang, X.; Gao, C.; Jiang, M.; Li, R.; Wang, Z.; Li, Y.; Fan, S.; Zhang, X. Effect of temperature, salinity and irradiance on growth and photosynthesis of *Ulva prolifera*. *Acta Oceanol. Sin.* **2016**, *35*, 114–121. [CrossRef]
24. Shimada, S.; Yokoyama, N.; Arai, S.; Hiraoka, M. Phylogeography of genus Ulva (Ulvophyceae, Chlorophyta), with special reference to the Japanese freshwater and brackish taxa. *J. Appl. Phycol.* **2008**, *20*, 979–989. [CrossRef]
25. Wu, H.; Gao, G.; Zhong, Z.; Li, X.; Xu, J. Physiological acclimation of the green tidal alga *Ulva prolifera* to a fast-changing environment. *Mar. Environ. Res.* **2018**, *137*, 1–7. [CrossRef]
26. Liu, D.; Keesing, J.K.; He, P.; Wang, Z.; Shi, Y.; Wang, Y. The world's largest macroalgal bloom in the Yellow Sea, China: Formation and implications. *Estuar. Coast. Shelf Sci.* **2013**, *129*, 2–10. [CrossRef]
27. Nigi, G.; Kinoshita, I.; Hiraoka, M.; Azuma, K. The effect of fluctuation in nutrients on the thallus length and pigment content of *Ulva prolifera* in the Shimanto River estuary. *Jpn. J. Phycol.* **2018**, *66*, 7–16.
28. Raven, J.A.; Geider, R.J. Temperature and algal growth. *New Phytol.* **1988**, *110*, 441–461. [CrossRef]
29. Valiela, I.; McClelland, J.; Hauxwell, J.; Behr, P.J.; Herch, D.; Foreman, K. Macroalgal blooms in shallow estuaries: Controls and ecophysiological and ecosystem consequences. *Limnol. Oceanogr.* **1997**, *42*, 1105–1118. [CrossRef]
30. Li, H.; Zhang, Y.; Chen, J.; Lie, F.; Jiao, N. Nitrogen uptake and assimilation preferences of the main green tide alga *Ulva prolifera* in the Yellow Sea, China. *J. Appl. Phycol.* **2019**, *31*, 625–635. [CrossRef]

31. Andersen, R.A.; Berges, J.A.; Harrison, P.J.; Watanabe, M.M. Appendix A—Recipes for freshwater and seawater media. In *Algal Culturing Techniques*; Andersen, R.A., Ed.; Elsevier Academic Press: Cambridge, UK, 2005; pp. 429–538.
32. Dan, A.; Hiraoka, M.; Ohno, M. On the induction of reproductive cell formation in the green alga, *Enteromorpha prolifera* Relationship of temperature and size of tissue disk. *Aquacult. Sci.* **1998**, *46*, 503–508, (In Japanese with English Abstract).

MDPI

Review

Massive *Ulva* Green Tides Caused by Inhibition of Biomass Allocation to Sporulation

Masanori Hiraoka 

Usa Marine Biological Institute, Kochi University, Inoshiri, Usa, Tosa, Kochi 781-1164, Japan; mhiraoka@kochi-u.ac.jp; Tel.: +81-88-856-0462

Abstract: The green seaweed *Ulva* spp. constitute major primary producers in marine coastal ecosystems. Some *Ulva* populations have declined in response to ocean warming, whereas others cause massive blooms as a floating form of large thalli mostly composed of uniform somatic cells even under high temperature conditions—a phenomenon called "green tide". Such differences in population responses can be attributed to the fate of cells between alternative courses, somatic cell division (vegetative growth), and sporic cell division (spore production). In the present review, I attempt to link natural population dynamics to the findings of physiological in vitro research. Consequently, it is elucidated that the inhibition of biomass allocation to sporulation is an important key property for *Ulva* to cause a huge green tide.

Keywords: biomass allocation; green tide; sporulation; *Ulva ohnoi*; *Ulva prolifera*; vegetative growth

Citation: Hiraoka, M. Massive *Ulva* Green Tides Caused by Inhibition of Biomass Allocation to Sporulation. *Plants* **2021**, *10*, 2482. https://doi.org/10.3390/plants10112482

Academic Editor: Koji Mikami

Received: 4 October 2021
Accepted: 13 November 2021
Published: 17 November 2021

Publisher's Note: MDPI stays neutral with regard to jurisdictional claims in published maps and institutional affiliations.

Copyright: © 2021 by the author. Licensee MDPI, Basel, Switzerland. This article is an open access article distributed under the terms and conditions of the Creative Commons Attribution (CC BY) license (https://creativecommons.org/licenses/by/4.0/).

1. Introduction

Ulva (Ulvophyceae, Chlorophyta) or sea lettuce is the most abundant green seaweed and is ubiquitous in tropical and temperate coastal ecosystems around the world. The genus *Ulva* currently includes at least 85 taxonomically accepted species [1]. The thallus body is uniformly sheet-like, being two cells thick or tubular with a single cell layer, except for a very small holdfast part (Figure 1A,B). The generation time of *Ulva* is short, and its spores can develop into a thallus having the potential ability to produce spores again in 2–3 weeks. Spore formation occurs in the somatic cells which directly transform into sporangia (Figure 1C,D). The sporulation first occurs at the tip of the algal body and then sequentially occurs toward the bottom, several tens of spores per sporangia are released, and the empty sporangium are spontaneously detached from the thallus.

A few *Ulva* species cause massive kilometer-scale blooms termed as green tides, and these have been recorded mainly around the industrialized coastlines of Europe, North America, and east Asia [2]. The world's most extensive *Ulva* green tides have repeatedly occurred in the Yellow Sea since 2007 [3] or 2008 [4]. The causative green tide species can substantially increase their biomass in a free-floating state by increasing the size of the thalli and their fragments. Smetacek and Zingone [5] pointed out that this is crucial because it is the unattached forms that, by invading new space, are able to increase their nutrient supply, free themselves from competition for limited hard substrates, and avoid their many benthic grazers. As a result, the unattached forms can build up a large biomass, forming massive green tides. In general, most *Ulva* spp. grow when their motile spores settle on the substrate. The attached populations in temperate regions are regularly present in early spring, increase to a maximum in late spring, and rapidly decrease through summer, showing a unimodal pattern of biomass fluctuation (Figure 2A) [6–9]. The summer decline of the attached populations occurs markedly at 20–25 °C. However, green tide species peak in biomass in the summer [10] and occasionally continue to grow [11]. Such a temporal difference of thallus growth pattern between attached populations and floating green tide populations has not received much attention in the literature. I consider that the species-specific differences of ecophysiological characteristics are a crucial key in determining if an algal species causes an excessive bloom or not.

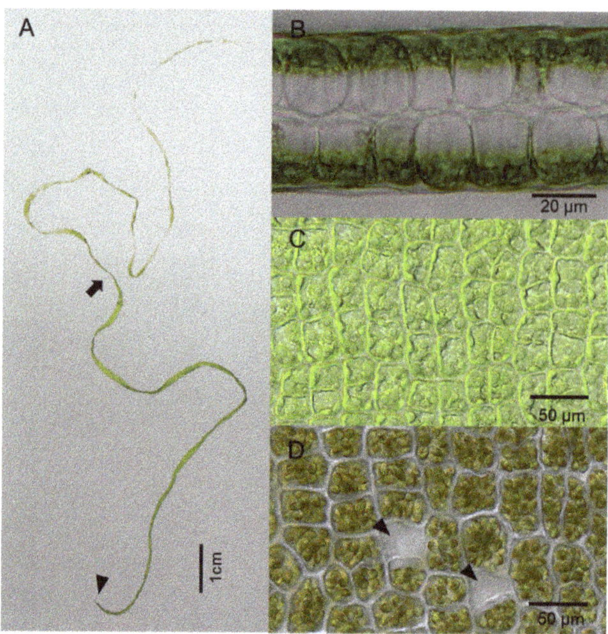

Figure 1. A living *Ulva* specimen (*U. aragoënsis*). (**A**) The developed thallus with sporulation in the upper part. Above the part indicated by the arrow, all the cells formed spores and some of them released spores. Arrowhead indicates a small holdfast; (**B**) Cross-section of the middle part of the thallus having a two-cell layered structure; (**C**) Surface view of somatic or blade cells in the vegetative state in the middle part of the thallus; (**D**) Surface view of the cells forming spores. Arrowheads indicate empty cells after spores were released.

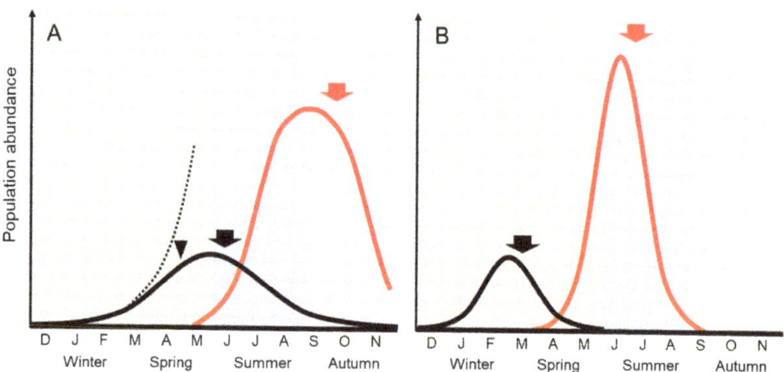

Figure 2. Comparison of seasonal abundance change between the attached populations and the green tides of *Ulva* in the temperate region. Arrowhead indicates the period when biomass increase is suppressed mainly by light limitation. Arrows show the period when allocation of biomass to sporulation begins to be greater than vegetative growth. (**A**) The attached *U. australis* modified from [9,12] (black line). Dotted line is a growth curve predicted in case of no light limitation. *Ulva ohnoi* green tide modified from [11] (red line); (**B**) Attached *U. prolifera* in brackish water modified from [13,14] (black line). The *U. prolifera* green tide in the Yellow Sea modified from [4,15] (red line).

In various previous publications, it has been explained that *Ulva* green tides are a symptom resulting from eutrophication [16]. Indeed, the supply of dissolved inorganic nitrogen (N) and phosphorus (P) is required to sustain *Ulva* thallus growth. However, as an example of green tides in Tokyo Bay, the scale of *Ulva* blooms has expanded despite a significant decrease in N and P concentrations year by year [17]. In this case, a subtropical species, *U. ohnoi*, unintentionally introduced and excessively grew as unattached form without the summer decline. In the Yellow Sea, free-floating *Ulva* populations during the early stage of green tides in spring include four or more *Ulva* species, but only one species, *U. prolifera*, continuously expand, resulting in monospecific spectacular blooms in summer [18,19]. All the other species do not seem to be able to continue vegetative growth while enduring the high temperatures from spring to summer, even though they sympatrically experience the same eutrophic conditions in the Yellow Sea. These examples of *U. ohnoi* and *U. prolifera* indicate that when specific species, having a continuous growth ability in high temperatures, encounter the minimum nutrient conditions for sustaining its vegetative growth, a huge bloom can occur.

In this review, I compare the *Ulva* biomass fluctuation patterns between attached populations and green tide populations and explain which ecophysiological properties of a species or strain is essential for causing green tides. In addition, I will focus on the mechanism of switching between the vegetative growth and sporulation of *Ulva* cells. By relating field observations and laboratory experiments, I attempt to more comprehensively examine the relationships between the cells, individuals, and populations underlying the mechanism of the development of green tides.

2. Attached Population Dynamics

2.1. Population Fluctuation Follows Individual Size Fluctuation

In the temperate coastal zone, the biomass of attached *Ulva* populations fluctuates seasonally according to the periodic fluctuation of the water temperature. Many temperate species such as *U. rigida*, *U. lactuca*, and *U. australis* (syn. *U. pertusa*) show a regular fluctuation pattern in which the attached biomass increases from winter to spring and declines during high temperatures from summer to autumn, as described in Figure 2A [6–9,12,20,21]. The biomass peaks shift to later in the year in cold regions at high latitudes [22]. This review will progress the story about the temperate populations. There has been detailed demographic research study on the attached population of *U. australis* over a period of three consecutive years [9]. It demonstrated that the seasonal fluctuations of the *Ulva* population synchronize with those of the thallus size rather than with changes in the density of thallus individuals. This indicates that biomass fluctuations of *Ulva* population are attributed to that of well-developed thallus individuals.

2.2. Increase Phase

Ulva with a simple multicellular body perform 'diffuse growth' in which cell divisions can occur more or less throughout the tissues of the organism [23]. The somatic cells divide synchronously in standardized conditions once a day [24]. Therefore, the *Ulva* thalli are capable of exponential growth, displaying extremely high growth rates. In fact, a daily rate of over fourfold in *U. meridionalis* in the culture experiment has been reported as the highest growth rate ever reported for multicellular autotrophic plants. In the same paper, a strain isolated from the attached *U. prolifera* population was also revealed to display two-fold growth rate per day [25]. The high exponential growth of *Ulva* spp. generally occurs in high temperatures of 20–30 °C in suitable light and nutrient conditions, as described below in Section 3. If such high growth continues in the sea from spring to summer with the optimum high temperatures, a bloom would occur explosively. However, in the attached population, the rapid biomass increase is suppressed mainly by light limitation caused by self-shading as density increases (Figure 2A). Although light limitation is an inevitable suppression factor, the population is also negatively affected by some irregular factors of low salinities by precipitation and herbivory by benthic organisms such as snails and sea hares. Their inhibitory effects occur because the population is attached to the substrate.

2.3. Decline Phase

As the water temperature rises over 20 °C, *Ulva* thalli are highly promoted to produce and release spores. The allocation to sporulation in thalli of the attached *Ulva* populations has been observed to be significantly larger during warmer months [12,20]. Niesenbaum [20] described "the seasonality of reproduction, and changes in the abundance of total biomass and reproductive biomass during the reproductive season could have been a function of temperature. The sharp decline of total biomass in early August, and its low rate of recovery through August and September were probably due to temperature effects on growth and reproduction". Furthermore as "When temperature reached seasonal highs, allocation of biomass to the formation and release of swarmers (spores) was greatest, while the rate of vegetative replacement diminished as temperatures first became suboptimal and then inhibitory for growth. This could explain the increases in percent reproductive tissue during August and September". These findings are essential for understanding the *Ulva* biomass fluctuation. However, so far the allocation of biomass to sporulation in the decline phase of *Ulva* populations has been almost overlooked. Practically, only the intrinsic traits involved in the increase phase, such as high growth rates or multiple reproduction modes, have been highlighted [3,26]. Recently, a few works examined the decline phase of the *U. prolifera* green tide in the Yellow Sea [10]. However, little coverage has been given to the allocation to sporulation.

2.4. Disappearance Phase

After the decline phase, *Ulva* thalli almost disappear in the autumn. In this disappearance phase, although evidence has not been provided from field research yet, small individual thalli of less than a few centimeters in size could release spores and disappear, whereby their settled spores grow up fast to small thalli and release spores again in the early developmental stage. This fast generation alternation may occur until the water temperature drops below 20 °C in temperate species. These predictions are derived from culture work in the laboratory as described next.

3. Individual Size Determined by Vegetative Growth and Sporulation Decay

Culture experiments using temperature-controlled incubators have confirmed that higher temperatures promote sporulation decay [27]. An asexual variant of *U. prolifera* originally isolated from the attached population in brackish water and its clonal offspring thalli were tested (Figure 3). According to this study, their growth rates increase as the temperature rises to 25 °C. However, sporulation at the apical part of the thallus occurs earlier as the temperature rises, and the amount of sporulation decay increases. At 30 °C after sporulation first occurs, the thalli repeatedly produce spores and as a whole continue to decrease, resulting in the disappearance in one and half months of culture. At 20 °C and 25 °C, the vegetative growth increment and the amount of decay due to sporulation are balanced, and the total length of the thalli cannot extend from about 10 cm. At 15 °C, because the vegetative growth exceeds the small amount of sporulation, the thalli continue to grow larger. At the low temperature of 10 °C, sporulation does not occur, and the thalli continue to grow slowly. From these findings, individual thallus mass (M) can be expressed by the two factors of vegetative growth (G) and the amount allocated to sporulation (S) as $M = G - S$. As the *U. prolifera* strain has the intrinsic trait of $S \geq G$ at ≥ 20 °C after the first sporulation, M becomes constant or decreases over 20 °C.

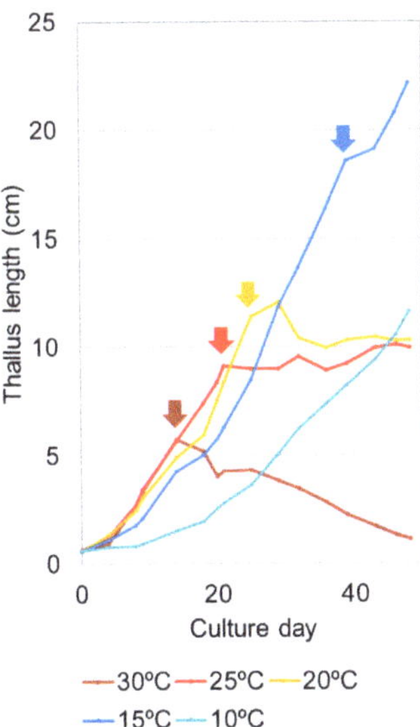

Figure 3. Change of averaged thallus length of *Ulva prolifera* strain isolated from the attached population in the Yoshino River estuary, Japan, cultured at different temperatures. Each arrow indicates the day when sporulation first occurred. After that, sporulation occurred repeatedly. Only at 10 °C, no sporulation occurred. This figure was redrawn based on the data from [27].

4. Inhibition of Sporulation Leads to Green Tides

4.1. *Ulva ohnoi* Green Tide

Floating thalli which are free from the attached substrate can spread moderately and receive sufficient light. If the floating thalli acquires the property of inhibiting sporulation even at high temperatures, they would make a massive green tide. Here, the intrinsic trait to cause a large-scale green tide is expressed as S < G at ≥20 °C. Then, as S is nearly zero and G is usually the maximum growth rate that the *Ulva* species can attain, M increases exponentially. In addition to the high-temperature growth property, the structural feature of easily producing floating thalli and their fragments also promotes the magnification of the green tides. *Ulva ohnoi* is a typical species with these characteristics. When this species was reported as a new green tide-forming species, it was taxonomically described as having a large, thin, and fragile blade thallus easily torn into floating fragments in the diagnosis [28]. The field observation of the *U. ohnoi* green tide was first made in Tosa Bay, southwestern Japan, for two years [11]. It shows that the thallus fragments grew rapidly at over a five-fold growth rate per week as the summer water temperature rose to 28–30 °C, reaching an average length of about 50 cm, a maximum length of >1 m, and the largest biomass attained approximately 1 kg fresh mass m^{-2} in August. In the decline phase of the green tides, a small amount of sporulation occurred in June-August, but half and more parts of the well-developed blades frequently formed and released spores in October, resulting in the population decline (Figure 2A). The remarkable difference compared to the temperate attached population is that *U. ohnoi* continues to grow with no or little biomass allocation to sporulation during the period of the summer decline. The properties are summarized below.

1. Sporulation which is generally promoted at high temperatures is suppressed.
2. At high temperatures, relatively high growth ability is exhibited.
3. Free-floating thallus fragments are easily formed.

Of these, 1 is particularly important for the occurrence of green tides. If the species does not produce spores and does not reduce its total mass, the green tide biomass gradually develops even if the growth is not so fast. A fragmentation culture method available for investigating the likelihood of sporulation in *Ulva* has been presented [29]. Applying the method, sporulation can be induced in 2 to 3 days on *Ulva* thallus tissue collected from the attached population [30]. By the same method, however, the thallus blades of *U. ohnoi* and the other *Ulva* spp. sampled from the massive green tides showed no or a very low frequency of induced sporulation or took a longer time to sporulation [31]. Before *U. ohnoi* was taxonomically differentiated as a new species, this species blooming in Ohmura Bay, southwestern Japan, had been identified as a sterile mutant of *U. pertusa* (now *U. australis*) [32] and it is still believed to be so [33]. Migita [32] showed that 1 cm^2 thallus fragments of his *U. ohnoi* strain displayed the maximum growth rate of two-fold growth rate per 2 days in laboratory experiments at 20 °C, and then transplanted into an outdoor tank and grew up to larger than 1 m^2 in 2 months without any sporulation, while all the fragments of more than 10 wild *U. australis* thalli formed spores in the same culture conditions, resulting in the disappearance of the thallus. These results indicate that bloom-forming species have a physiological property of being less prone to sporulate, or they do not sporulate.

Ulva ohnoi distribute mainly in the subtropical region and are adapted to high temperatures. Therefore, it has spread to the temperate area and outbreaks in the summer. The spread of *U. ohnoi* has been increasingly reported from various regions [34–36]. Due to global warming, *U. ohnoi* may spread further into higher latitudes and cause green tides. However, in Tosa Bay, where the *U. ohnoi* green tide was first reported, this species has recently decreased sharply and instead, *U. reticulata*, which has a distribution centered in more tropical waters, has begun to increase [37]. This example suggests that each *Ulva* species has a temperature range that balances the vegetative growth and sporulation, and that individual thallus growth, or population growth, may not be possible if the temperature limit is exceeded even by a few degrees.

4.2. *Ulva prolifera* Green Tide

The *U. prolifera* green tides in the Yellow Sea regularly reach their biomass peak during June and July in summer (Figure 2B) [4,15]. The earliest free-floating *Ulva* patches are found in the coastal areas of the southern Yellow Sea from mid-April to early May [3,18]. These patches originally appear off the nearby rafts for purple laver (*Neopyropia yezoensis*) aquaculture, for which the coverage area is approximately 4.1 × 10^4 ha [3]. Annually, approximately 6500 t of the *Ulva* mass has been estimated to be released as macroalgal waste from mid-April to late-May after cleaning the *Neopyropia* aquaculture facilities [38]. In this early stage, the patches include multiple *Ulva* spp. such as *U. linza*, *U. compressa*, and *U. aragoënsis* (=*U. flexuosa* in [18,19]). However, the free-floating *Ulva* complexes move northward, associated with the seasonal monsoons and ocean currents, rapidly develop into long large bands ranging from hundreds of meters to tens of kilometers in the open sea area in late May [3], and then massive green tides are dominated by a single species, *U. prolifera* [18,19,39]. Only this species explosively grows, while the other species disappear over 20 °C in early summer. This suggests that the bloom-forming *U. prolifera* can continuously grow inhibiting allocation to sporulation in high temperatures, thereby differing from the other *Ulva* spp. in this region. Supporting this finding, the culture work showed that a few centimeters of the bloom-forming *U. prolifera* fragment can grow to more than 50 cm in length without sporulation at 20 °C (cf. Figure 12 in [40]). This growth characteristic is obviously different from that of the *U. prolifera* strain from the attached population, which cannot grow over 10 cm at ≥20 °C (Figure 3). However, the bloom-forming *U. prolifera* seems to allocate its biomass to sporulation around 25 °C, because it was observed that the green tide population began to decline at 25 °C from July to August [10].

Different from the bloom-forming type, the common attached type of *U. prolifera* form abundant populations on the substrate in brackish waters such as river estuaries [41,42]. Seasonal fluctuations of the largest attached *U. prolifera* population in Japan have been described in detail (Figure 2B) [13,14]. The biomass and thallus length of the attached population regularly reach their maximum from January to March and then disappear by July. Although natural *U. prolifera* mats develop in the Chinese coast located in the south of the Yellow Sea and are harvested as edible biomass, the peak harvest is also from January to March [43], which is consistent with the fluctuation pattern of the populations in Japan. Although the attached *U. prolifera* has been used as an expensive macroalgal ingredient for Japanese dishes for a long time, its harvest has become a 'winter' tradition [13]. However, in line with recent ocean warming, the annual yield is declining [44]. It can be explained that the biomass decrease of the attached population is due to the shortening of the period when the water temperature falls below 20 °C. The most significant difference of the ecophysiological property between the bloom-forming type and the attached type in *U. prolifera* is whether they can vegetatively grow inhibiting allocation to sporulation around 20 °C or not. The seasonal difference is described in Figure 4.

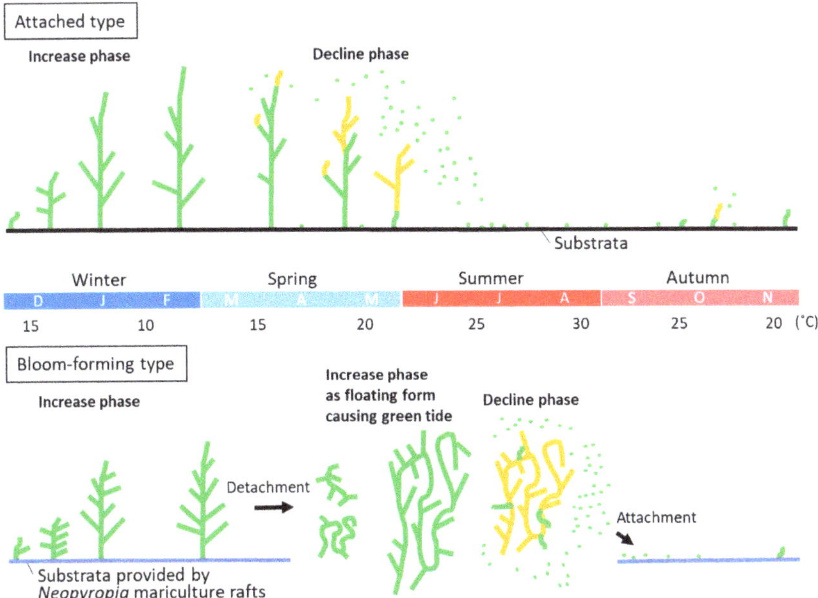

Figure 4. Comparison of seasonal change of thallus state between the attached type, *Ulva prolifera* subsp. *prolifera*, and the bloom-forming type, *U. prolifera* subsp. *qingdaoensis*. Green and orange lines of thallus image show vegetative state and sporulating state, respectively. Green dots indicate microscopic propagules or spores released from the sporulated thalli. The two types have significantly different timings of the decline phase.

Although the issue of the massive green tide in China received a great deal of attention in 2008 [45], immediately after that in some research papers the bloom-forming type and the attached type were not distinguished, and both were confused because they formed a monophyletic clade together with the most closely related species, *U. linza*, by molecular analysis using the nuclear-encoded rDNA internal transcribed spacer (ITS) region, which is commonly used for *Ulva* species identification. However, studies of culture, hybridization, and phylogenetic analysis using a higher resolution DNA marker (5S rDNA spacer region) revealed that the bloom-forming type can cross with the attached type without any reproductive boundary, which was confirmed to be conspecific, but there are several genetic and ecophysiological differentiations (Table 1). Consistently, the other genetic analyses using inter-simple sequence repeat markers and a sequence-characterized amplified region marker indicated that the bloom-forming type is a unique ecotype of *U. prolifera*, genetically distinct from the attached types along the Chinese coast [46].

Table 1. Comparison between the bloom-forming type and the attached type in *Ulva prolifera*.

	Bloom-Forming Type	Attached Type
Subspecies name	*Ulva prolifera* subsp. *qingdaoensis*	*Ulva prolifera* subsp. *prolifera*
Habitat/Distribution	Offshore and coastal waters in the Yellow Sea	Brackish and estuarine waters in cold–temperate regions
Season for maximum population abundance	Early summer	Winter to early spring
Temperature at which thallus growth turns to zero or negative	Around 25 °C	Around 20 °C
Branch density of cultured thalli	>100 per 1 cm of main stem	0.4–14 per 1 cm of main stem [1]
Life cycle	Sexual	Sexual or obligate asexual by special spore (zoosporoid)
Crossing affinity to *Ulva linza*	Complete incompatibility	Partial gamete incompatibility
5S sequence	5S-A type or 5S-B type [2]	A large number of various types [3] different from 5S-A and 5S-B types

[1] Data from the type locality population [40]; [2] The two types reported by [18]; [3] Thirty-one types reported by [42] and several more [19,43,47].

From the phylogenetic analyses of *U. linza* and *U. prolifera* using the 5S sequence, it has been suggested that *U. prolifera* had adapted to brackish water and recently evolutionarily separated from marine *U. linza* [42]. Furthermore, crossing tests suggested that the bloom-forming type of *U. prolifera* completed the speciation from *U. linza* via intermediate brackish *U. prolifera* (=the attached type) because there is still a partial compatibility between the common brackish *U. prolifera* and *U. linza*, but a complete reproductive barrier exists between the bloom-forming *U. prolifera* and *U. linza* [48]. The brackish *U. prolifera* contains many regional populations that have genetically differentiated [42]. One of them may have acquired the ability to suppress sporulation and continue vegetative growth even at 20 °C or higher, resulting in creating the *U. prolifera* subsp. *qingdaoensis* that cause green tides. Interestingly, the *U. prolifera* subsp. *qingdaoensis* is characterized by a densely branching morphology (Table 1). This feature may facilitate the production and dispersal of large numbers of floating thallus fragments.

In contrast to *U. ohnoi*, widely reported from various regions, the occurrence of *U. prolifera* green tides has been limited to the Yellow Sea only. The special *U. prolifera* population unique to this region is fostered in the vast *Neopyropia* farm as its nursery bed, and is supplied annually as a large amount of floating mass [3,38]. The world's largest green tide appears to be caused by such a very special production cycle supported by the aquaculture activities.

5. Mechanism of Sporulation

The multicellular body of *Ulva* is composed of mostly uniform blade cells (or somatic cells) except for a small number of rhizoid cells forming a small holdfast (Figure 1). Therefore, the allocation of the individual thallus tissue to vegetative growth and sporulation is almost attributed to the blade cell fate between the alternative courses, somatic cell division, and sporic cell division. Nordby [29] hypothesized that the cell fate is controlled by changes in the concentration of sporulation inhibitors. This inhibitor hypothesis inferred that the double-layered structure of the *Ulva* thallus (Figure 1B) could be responsible for maintaining a sufficient concentration of the sporulation inhibitor during vegetative growth. Supporting that, Stratmann et al. [49] revealed that the *Ulva* thallus produces at least two kinds of the sporulation inhibitor, one of which is a glycoprotein 'Sporulation inhibitor-1′ (SI-1), and the other is a nonprotein of very low molecular mass (SI-2). The SI-1 is present in the cell wall of *Ulva* cells and appears to be secreted extracellularly. The SI-2 is in the inner space between the two blade cell layers. The excretion of the SI-1 decreases with maturation of the thallus, whereas the overall concentration of SI-2 in the thallus stays constant throughout the life cycle. The SI-2 affects different *Ulva* species whereas the SI-1 is species-specific. Although such characteristics of the inhibitors have been shown, their molecular structures have not yet been identified.

The fragmentation culture method can synchronously induce sporulation in *Ulva* thallus tissue within 48 h, when fragmented single-layered thalli are transferred to fresh medium at a low density of fragments and cultured in optimal conditions [29]. It is explained by the inhibitor hypothesis that the inhibitors leak out from the circumference of fragmented thalli and the somatic cells that sense the decrease in the concentration of the inhibitors go to sporulation. As already mentioned, the bloom-forming *Ulva* spp. hardly, or do not, allocate the somatic cells to sporulation, even in high temperatures. Considering the inhibitor hypothesis, it is possibly thought that the bloom-forming species produce a large amount of the inhibitory substances or have an inhibitor-sensing system in which it is hardly relieved from the inhibition. The entire genome of *Ulva* has already been announced [50]. Therefore, if the inhibitors are structurally identified, it is expected that the elucidation of the switching mechanism between the somatic cell division and the sporic cell division will be achieved.

6. Conclusions and Perspective

By comparing attached populations and green tide populations, it became clear that the bloom-forming species continue to grow vegetatively with almost no spore formation even at high temperatures. However, few field surveys have been conducted on the process of the decline of green tide populations from the viewpoint of the allocation to sporulation, and future investigations are required.

Though the sporulation inhibitors were partially characterized in 1996, their molecular structures have not yet been determined. These substances are the key to determining the fate of somatic cell division or sporic cell division. It is highly possible that the causative species of the green tide have different reaction systems involving the sporulation inhibitors. Such research studies are more likely to detect minor differences when comparing taxa containing blooming strains and non-blooming strains within the same species. In that sense, *U. prolifera* would be excellent experimental material.

Ulva is a promising organism for carbon dioxide fixation and bioproduct production due to its high productivity [51,52]. Understanding the allocation system of vegetative growth and sporulation decay will enable greater control of biomass production and will contribute to the development of the bioeconomy.

Funding: This research was supported by the Kochi University research project of the Biomass Refinery of Marine Algae, the JST-OPERA Program (Grant Number JPMJOP1832), and the Wood and Cabinet Office grant in aid, the Advanced Next-Generation Greenhouse Horticulture by IoP (Internet of Plants), Japan.

Conflicts of Interest: The author declares no conflict of interest.

References

1. Guiry, M.D.; Guiry, G.M. *AlgaeBase*; World-Wide Electronic Publication; National University of Ireland: Galway, Ireland; Available online: http://www.algaebase.org (accessed on 4 September 2021).
2. Joniver, C.F.H.; Photiades, A.; Moore, P.J.; Winters, A.L.; Woolmer, A.; Adams, J.M.M. The global problem of nuisance macroalgal blooms and pathways to its use in the circular economy. *Algal Res.* **2021**, *58*, 102407. [CrossRef]
3. Zhang, Y.; He, P.; Li, H.; Li, G.; Liu, J.; Jiao, F.; Zhang, J.; Huo, Y.; Shi, X.; Su, R. *Ulva prolifera* green-tide outbreaks and their environmental impact in the Yellow Sea, China. *Nat. Sci. Rev.* **2019**, *6*, 825–838. [CrossRef]

4. Qi, L.; Hu, C.; Xing, Q.; Shang, S. Long-term trend of *Ulva prolifera* blooms in the western Yellow Sea. *Harmful Algae* **2016**, *58*, 35–44. [CrossRef] [PubMed]
5. Smetacek, V.; Zingone, A. Green and golden seaweed tides on the rise. *Nature* **2013**, *504*, 84–88. [CrossRef] [PubMed]
6. Rivers, J.S.; Peckol, P. Summer decline of *Ulva lactuca* (Chlorophyta) in a eutrophic embayment: Interactive effects of temperature and nitrogen availability? *J. Phycol.* **1995**, *31*, 223–228. [CrossRef]
7. Choi, T.S.; Kim, J.H.; Kim, K.Y. Seasonal changes in the abundance of *Ulva* mats on a rocky intertidal zone of the southern coast of Korea. *Algae* **2001**, *16*, 337–341.
8. Kim, K.Y.; Choi, T.S.; Kim, J.H.; Han, T.; Shin, H.W.; Garbary, D.J. Physiological ecology and seasonality of *Ulva pertusa* on a temperate rocky shore. *Phycologia* **2004**, *43*, 483–492. [CrossRef]
9. Hiraoka, M.; Yoshida, G. Temporal variation in isomorphic phase and sex ratios of a natural population of *Ulva pertusa* (Chlorophyta). *J. Phycol.* **2010**, *46*, 882–888. [CrossRef]
10. Hao, Y.; Guan, C.; Zhao, X.; Qu, T.; Tang, X.; Wang, Y. Investigation of the decline of *Ulva prolifera* in the Subei Shoal and Qingdao based on physiological changes. *J. Ocean. Limnol.* **2021**, *39*, 918–927. [CrossRef]
11. Ohno, M. Seasonal changes of the growth of green algae, *Ulva* sp. in Tosa Bay, southern Japan. *Mar. Fouling* **1988**, *7*, 13–17. [CrossRef]
12. Uchimura, M.; Yoshida, G.; Hiraoka, M.; Komatsu, T.; Arai, S.; Terawaki, T. Ecological studies of green tide, *Ulva* spp. (Chlorophyta) in Hiroshima Bay, the Seto Inland Sea. *Jpn. J. Phycol.* **2004**, *52*, 17–22.
13. Ohno, M.; Mizutani, S.; Taino, S.; Takahashi, I. Ecology of the edible green alga *Enteromorpha prolifera* in Shimanto River, southern Japan. *Bull. Mar. Sci. Fish. Kochi Univ.* **1999**, *19*, 27–35.
14. Hiraoka, M.; Higa, M. Novel distribution pattern between coexisting sexual and obligate asexual variants of the true estuarine macroalga *Ulva prolifera*. *Ecol. Evol.* **2016**, *6*, 3658–3671. [CrossRef] [PubMed]
15. Hu, L.; Zeng, K.; Hu, C.; He, M.X. On the remote estimation of *Ulva prolifera* areal coverage and biomass. *Remote Sens. Environ.* **2019**, *223*, 194–207. [CrossRef]
16. Teichberg, M.; Fox, S.E.; Olsen, Y.S.; Valiela, I.; Martinetto, P.; Iribarne, O.; Muto, E.; Petti, M.A.V.; Corbisier, T.N.; Soto-Jiménez, M.; et al. Eutrophication and macroalgal blooms in temperate and tropical coastal waters: Nutrient enrichment experiments with *Ulva* spp. *Glob. Chang. Biol.* **2010**, *16*, 2624–2637. [CrossRef]
17. Yabe, T.; Ishii, Y.; Amano, Y.; Koga, T.; Hayashi, S.; Nohara, S.; Tatsumoto, H. Green tide formed by free-floating *Ulva* spp. at Yatsu tidal flat, Japan. *Limnology* **2009**, *10*, 239–245. [CrossRef]
18. Han, W.; Chen, L.P.; Zhang, J.H.; Tian, X.L.; Hua, L.; He, Q.; Huo, Y.Z.; Yu, K.F.; Shi, D.J.; Ma, J.H.; et al. Seasonal variation of dominant free-floating and attached *Ulva* species in Rudong coastal area, China. *Harmful Algae* **2013**, *28*, 46–54. [CrossRef]
19. Zhang, Q.C.; Yu, R.C.; Chen, Z.F.; Qiu, L.M.; Wang, Y.F.; Kong, F.Z.; Geng, H.X.; Zhao, Y.; Jiang, P.; Yan, T.; et al. Genetic evidence in tracking the origin of *Ulva prolifera* blooms in the Yellow Sea, China. *Harmful Algae* **2018**, *78*, 86–94. [CrossRef]
20. Niesenbaum, R.A. The ecology of sporulation by the macroalga *Ulva lactuca* L. (Chlorophyceae). *Aquat. Bot.* **1988**, *32*, 155–166. [CrossRef]
21. Fillit, M. Seasonal changes in the photosynthetic capacities and pigment content of *Ulva rigida* in a Mediterranean coastal lagoon. *Bot. Mar.* **1995**, *38*, 271–280. [CrossRef]
22. Munda, I.M.; Markham, J.W. Seasonal variations of vegetation patterns and biomass constituents in the rocky eulittoral of Helgoland. *Helgoländer Meeresunters* **1982**, *35*, 131–151. [CrossRef]
23. Garbary, D.; Galway, M.E. Macroalgae as underexploited model systems for stem cell research. In *Deferring Development*; Bishop, C.D., Hall, B.K., Eds.; CRC Press: New York, NY, USA, 2020; pp. 87–105.
24. Wichard, T.; Charrier, B.; Mineur, F.; Bothwell, J.H.; De Clerck, O.; Coates, J.C. The green seaweed *Ulva*: A model system to study morphogenesis. *Front. Plant Sci.* **2015**, *6*, 72. [CrossRef]
25. Hiraoka, M.; Kinoshita, Y.; Higa, M.; Tsubaki, S.; Monotilla, A.P.; Onda, A.; Dan, A. Fourfold daily growth rate in multicellular marine alga *Ulva meridionalis*. *Sci. Rep.* **2020**, *10*, 12606. [CrossRef] [PubMed]
26. Fort, A.; Mannion, C.; Fariñas-Franco, J.M.; Sulpice, R. Green tides select for fast expanding *Ulva* strains. *Sci. Total Environ.* **2020**, *698*, 134337. [CrossRef] [PubMed]
27. Hiraoka, M.; Dan, A.; Hagihira, M.; Ohno, M. Growth and maturity of clonal thalli in *Enteromorpha prolifera* under different temperature conditions. *Nippon Suisan Gakkaishi* **1999**, *65*, 302–303. [CrossRef]
28. Hiraoka, M.; Shimada, S.; Uenosono, M.; Masuda, M. A new green-tide-forming alga, *Ulva ohnoi* Hiraoka et Shimada sp. nov. (Ulvales, Ulvophyceae) from Japan. *Phycol. Res.* **2004**, *52*, 17–29. [CrossRef]
29. Nordby, Ø. Optimal conditions for meiotic spore formation in *Ulva mutabilis* Føyn. *Bot. Mar.* **1977**, *20*, 19–28. [CrossRef]
30. Hiraoka, M.; Enomoto, S. The induction of reproductive cell formation of *Ulva pertusa* Kjellman (Ulvales, Ulvophyceae). *Phycol. Res.* **1998**, *46*, 199–203. [CrossRef]
31. Hiraoka, M.; Ohno, M.; Kawaguchi, S.; Yoshida, G. Crossing test among floating *Ulva* thalli forming 'green tide' in Japan. *Hydrobiologia* **2004**, *512*, 239–245. [CrossRef]
32. Migita, S. The sterile mutant of *Ulva pertusa* Kjellman from Oumra Bay. *Bull. Fac. Fish. Nagasaki Univ.* **1985**, *57*, 33–37.
33. Gao, G.; Clare, A.S.; Rose, C.; Caldwella, G.S. Reproductive sterility increases the capacity to exploit the green seaweed *Ulva rigida* for commercial applications. *Algal Res.* **2017**, *24*, 64–71. [CrossRef]

34. Melton, J.T., III; Collado-Vides, L.; Lopez-Bautista, J.M. Molecular identification and nutrient analysis of the green tide species *Ulva ohnoi* M. Hiraoka & S. Shimada, 2004 (Ulvophyceae, Chlorophyta), a new report and likely nonnative species in the Gulf of Mexico and Atlantic Florida, USA. *Aquat. Invasions* **2016**, *11*, 225–237.
35. Krupnik, N.; Rinkevich, B.; Paz, G.; Douek, J.; Lewinsohn, E.; Israel, A.; Carmel, N.; Mineur, F.; Maggs, C.A. Native, invasive and cryptogenic *Ulva* species from the Israeli Mediterranean Sea: Risk and potential. *Mediterr. Mar. Sci.* **2018**, *19*, 132–146. [CrossRef]
36. Chávez-Sánchez, T.; Piñón-Gimate, A.; Melton, J.T., III; López-Bautista, J.M.; Casas-Valdez, M. First report, along with nomenclature adjustments, of *Ulva ohnoi, U. tepida* and *U. torta* (Ulvaceae, Ulvales, Chlorophyta) from northwestern Mexico. *Bot. Mar.* **2019**, *62*, 113–123. [CrossRef]
37. Hiraoka, M.; Tanaka, K.; Yamasaki, T.; Miura, O. Replacement of *Ulva ohnoi* in the type locality under rapid ocean warming in southwestern Japan. *J. Appl. Phycol.* **2020**, *32*, 2489–2494. [CrossRef]
38. Wang, Z.; Xiao, J.; Fan, S.; Li, Y.; Liu, X.; Liu, D. Who made the world's largest green tide in China?—An integrated study on the initiation and early development of the green tide in Yellow Sea. *Limnol. Oceanogr.* **2015**, *60*, 1105–1117. [CrossRef]
39. Wang, S.; Huo, Y.; Zhang, J.; Cui, J.; Wang, Y.; Yang, L.; Zhou, Q.; Lu, Y.; Yu, K.; He, P. Variations of dominant free-floating *Ulva* species in the source area for the world's largest macroalgal blooms, China: Differences of ecological tolerance. *Harmful Algae* **2018**, *74*, 58–66. [CrossRef] [PubMed]
40. Cui, J.; Monotilla, A.P.; Zhu, W.; Takano, Y.; Shimada, S.; Ichihara, K.; Matsui, T.; He, P.; Hiraoka, M. Taxonomic reassessment of *Ulva prolifera* (Ulvophyceae, Chlorophyta) based on specimens from the type locality and Yellow Sea green tides. *Phycologia* **2018**, *57*, 692–704. [CrossRef]
41. Hiraoka, M.; Dan, A.; Shimada, S.; Hagihira, M.; Migita, M.; Ohno, M. Different life histories of *Enteromorpha prolifera* (Ulvales, Chlorophyta) from four rivers on Shikoku Island Japan. *Phycologia* **2003**, *42*, 275–284. [CrossRef]
42. Shimada, S.; Yokoyama, N.; Arai, S.; Hiraoka, M. Phylogeography of the genus *Ulva* (Ulvophyceae, Chlorophyta), with special reference to the Japanese freshwater and brackish taxa. *J. Appl. Phycol.* **2008**, *20*, 979–989. [CrossRef]
43. Liu, F.; Pang, S.; Zhao, X. Molecular phylogenetic analyses of *Ulva* (Chlorophyta, Ulvophyceae) mats in the Xiangshan Bay of China using high-resolution DNA markers. *J. Appl. Phycol.* **2013**, *25*, 1287–1295. [CrossRef]
44. Hiraoka, M. Seaweed bed shifts under enhanced ocean warming and measures to cover fishery losses. *Aquabiology* **2012**, *34*, 314–318.
45. Leliaert, F.; Zhang, X.; Ye, N.; Malta, E.; Engelen, A.H.; Mineur, F.; Verbruggen, H.; De Clerck, O. Identity of the Qingdao algal bloom. *Phycol. Res.* **2009**, *57*, 147–151. [CrossRef]
46. Zhao, J.; Jiang, P.; Qin, S.; Liu, X.; Liu, Z.; Lin, H.; Li, F.; Chen, H.; Wu, C. Genetic analyses of floating *Ulva prolifera* in the Yellow Sea suggest a unique ecotype. *Estuar. Coast. Shelf Sci.* **2015**, *163*, 96–102. [CrossRef]
47. Duan, W.; Guo, L.; Sun, D.; Zhu, S.; Chen, X.; Zhu, W.; Xu, T.; Chen, C. Morphological and molecular characterization of free-floating and attached green macroalgae *Ulva* spp. in the Yellow Sea of China. *J. Appl. Phycol.* **2012**, *24*, 97–108. [CrossRef]
48. Hiraoka, M.; Ichihara, K.; Zhu, W.; Ma, J.; Shimada, S. Culture and hybridization experiments on an *Ulva* clade including the Qingdao strain blooming in the Yellow Sea. *PLoS ONE* **2011**, *6*, e19371. [CrossRef]
49. Stratmann, J.; Paputsoglu, G.; Oertel, W. Differentiation of *Ulva mutabilis* (Chlorophyta) gametangia and gamete release are controlled by extracellular inhibitors. *J. Phycol.* **1996**, *32*, 1009–1021. [CrossRef]
50. De Clerck, O.; Kao, S.M.; Bogaert, K.A.; Blomme, J.; Foflonker, F.; Kwantes, M.; Vancaester, E.; Vanderstraeten, L.; Aydogdu, E.; Boesger, J.; et al. Insights into the evolution of multicellularity from the sea lettuce genome. *Curr. Biol.* **2018**, *28*, 2921–2933.e5. [CrossRef] [PubMed]
51. Lawton, R.J.; Sutherland, J.E.; Glasson, C.R.; Magnusson, M.E. Selection of temperate *Ulva* species and cultivars for land-based cultivation and biomass applications. *Algal Res.* **2021**, *56*, 102320. [CrossRef]
52. Moreira, A.; Cruz, S.; Marques, R.; Cartaxana, P. The underexplored potential of green macroalgae in aquaculture. *Rev. Aquac.* **2021**. [CrossRef]

MDPI
St. Alban-Anlage 66
4052 Basel
Switzerland
Tel. +41 61 683 77 34
Fax +41 61 302 89 18
www.mdpi.com

Plants Editorial Office
E-mail: plants@mdpi.com
www.mdpi.com/journal/plants

www.ingramcontent.com/pod-product-compliance
Lightning Source LLC
LaVergne TN
LVHW070610170726
843515LV00004B/29

* 9 7 8 3 0 3 6 5 3 3 9 2 6 *